The Algorithm Writer's Guide

D. M. Wheatley and
A. W. Unwin

THE ALGORITHM WRITER'S GUIDE

Longman

LONGMAN GROUP LIMITED
London
Associated companies, branches and representatives throughout the world

© Longman Group Limited 1972

First published 1972

ISBN 0 582 42161·6 cased
ISBN 0 582 42162·4 paper

Set by Filmtype Services Limited, Scarborough and printed in Great Britain by
Lowe & Brydone (Printers) Ltd., Thetford, Norfolk

CONTENTS

INTRODUCTION

The algorithm is already well established as a means of plotting the strategy of problem solving in computers. Its use as a tool which enables *people* to solve problems is more recent, but has been gaining ground over the past few years in industry, commerce, government and the services. Algorithms are now being used to interpret regulations and laws, to detect faults in machinery and electronic systems, to trace the causes of defects in products and for many other purposes.

Previous writings on algorithms have been either for the specialist, who wants to know the latest research, or for the beginner, who needs to understand how algorithms work and what use they can be to him. In both cases, the explanation given has been in terms of a theoretical model – the idea of the algorithm itself – which has been applied to practical problems.

The present book approaches largely from the opposite direction. The variations in form and method it describes arise directly from the work of Cambridge Consultants (Training) Ltd., and most of the examples were produced to solve specific problems brought to us by our clients. We frequently found that, although it would have been possible to use 'standard' flow-chart algorithms, the best solution to the problem involved some variation in layout or design. The one criterion in day-to-day consultancy work is that what we produce should be worth the money our clients pay for it, and it was sometimes only after the job was finished that we realised the implications of what we had done, and generalised our solution into a method that could be applied to other situations. .

Our approach, therefore, is entirely practical. What we seek to achieve is to give the writer, or would-be writer, of algorithms the information he needs to do his job. We have tried to do this in three ways: firstly by giving and analysing examples of many different types of algorithm; secondly by guiding the reader through the process of writing specific algorithms; thirdly by describing various techniques which may be of value in coping with special problems.

We do not claim that everyone who reads this book will be able to write effective algorithms. A certain analytical ability and a good deal of practice are also required. What we claim is that anyone who sets out to write algorithms will have a greater chance of success if he has some feeling for what is involved, some knowledge of the techniques developed so far, and an acquaintance with the many possible forms an algorithm can take.

The obvious omission from this list is the 'principles of algorithm writing'. The omission is no accident. While we have devoted considerable space to discussion of matters of principle, we have deliberately left the edges of the subject a little blurred, for our aim is not definition but development.

1 WHAT IS AN ALGORITHM?

The purpose of this chapter is to give those who wish to write algorithms a general view of the types of algorithm that have been developed so far, and to discuss various attempts to define what an algorithm is or should be.

Present uses of the word seem to be derived from theoretical mathematics, where it means 'an exact prescription defining a computational process that leads from various initial data to a desired result' (Markov, 1961). It is, in fact, a mathematical recipe. From this its meaning has been extended to cover a recipe in any field of activity. An analogy is drawn with computer programming, where a computer is made to work through a set of procedures, i.e. an algorithm, to achieve a desired result. Similarly, it is argued, people can work through algorithms automatically and are led to a decision, the solution of a problem, the location of a fault in a piece of machinery, or whatever the algorithm is about. No wider understanding of the problem is necessary in order to reach the solution.

That they can be used to solve problems is undoubtedly the case. The algorithm below, which is laid out in the usual flow-chart format, is part of a sequence covering the law of adoption.

Two persons wish to make a joint application to adopt a child.

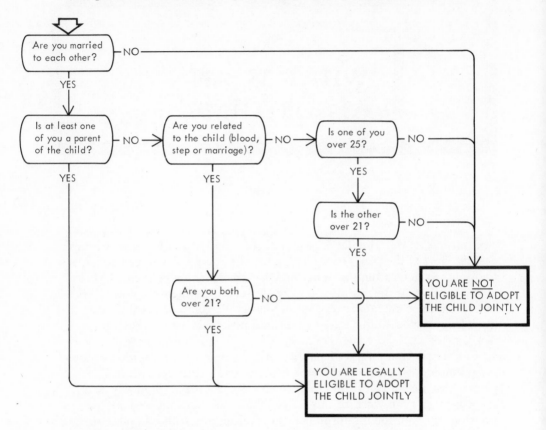

Let us suppose that a man aged 24 and his wife aged 22 wish to adopt the wife's niece. Are they eligible in law to do so? Taking the questions they would answer in order:

1. They *are* married to each other (Follow the YES-line).
2. They are *not* parents of the child.
3. They *are* related to the child.
4. They *are* both over 21.

They are therefore legally eligible to adopt the child. This we found out in four simple steps; if we had used conventional legal literature, it would certainly have taken very much longer.

Algorithms of this non-mathematical type have been described as 'a sequence of simple sentences (or questions) ordered in a logical hierarchy from the most general to the most specific, in such a way that only those sentences need be read which are relevant to a particular case' (Wason 1962).

This last point concerning the relevance of the sentences or questions can be well illustrated if we see what happens when we transform a rule into an algorithm. The following, we will suppose, is taken from the regulations of a club:

RULE 5: A man pays an annual subscription of £5, unless he is under 25 years of age, when it is reduced by half. A woman pays a subscription of £4, unless she is under 21 years of age, when it is reduced by half, except that when she is married she may pay the reduced subscription irrespective of age.

Now supposing a married woman aged 25 wants to know what subscription she should pay. How much of Rule 5, as it now stands, will she have to read to find this out? The answer, of course, is all of it. Even though she would have not the slightest interest in the subscriptions paid by men over 25, men under 25, unmarried women over 21 and women under 21, she would still have to read all about them to find the answer to her own case. Of course, the chances are she would turn from the rule book with a shudder and ask the Hon. Treasurer instead, who, after a quick glance at her wedding ring, would sign her up for £2. It is easy for the Hon. Treasurer. He has the algorithm in his head. The problem is to get it on to paper for everyone to use.

The first line of Rule 5 presents no problem:

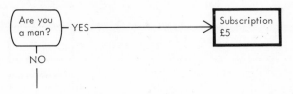

(The 'NO' will, of course, lead to the regulations for women's subscriptions.)

The second line forces us to qualify our algorithm a little. Men, we read, only pay £5 if they are over 25. If they are under 25 they pay half this.

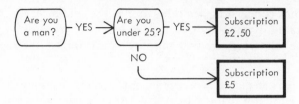

So much for the men. Rule 5 specifies similar conditions for women: if under 21 they pay half the subscription; if over 21 they pay the full £4.

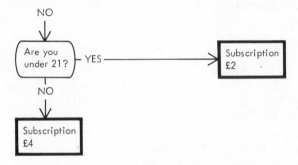

But the subscription is also halved for married women, whatever their age. Since this has the same effect as being under 21 the question can go in the same box.

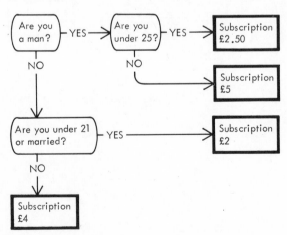

Now we can see how the Hon. Treasurer gets his answers so quickly. First: 'Are you a man?' He concludes that she is not. Second: 'Are you under 21 or married?' On the strength of the wedding ring he concludes that she is, and charges her £2.

Clearly, therefore, the idea of 'reading only those sentences which are relevant to a particular case' points to one of the essential features of algorithms. The rest of the definition is more problematic. Why should the sentences or questions necessarily proceed 'from the most general to the most specific'? If all cases occur with equal frequency this is reasonable enough, but what if one particular case occurs 90 per cent of the time? For example, the logically correct way of tracing a fault on a car's rearlights might be to start at the battery (the most general) and trace the wiring through to the rearlight (the most specific). The electrician who usually does this job, however, might know that this particular vehicle is prone to earthing faults. What he knows in effect is that if he always starts with a quick check for bad earth connections he will usually save himself a lot of time compared with someone who carries out the full logical procedure. If he is correct in thinking this, the most practical algorithm would be one which started with a quite specific point – an instruction to make a quick check of all relevant earth connections – before going on to the more general checks.

Another case in which the precision of a definition appears to outstrip its usefulness is the distinction that is made between algorithms and logical trees. Logical trees, we are told, 'consist solely of linked questions (with no "instructions" intervening) each of which helps to guide the reader to an unambiguous solution. A logical tree is a special kind of algorithm ideally suited for communicating rules and regulations.' (Lewis and Woolfenden) According to this definition, the algorithms on adoption and Rule 5 should really be called logical trees, since they contain questions only, with no instructions intervening. Such instructions are especially likely to occur in algorithms dealing with mechanical procedures, such as the following which is part of a sequence on electrical faults in cars.

Starter is not turning and lights do not dim.

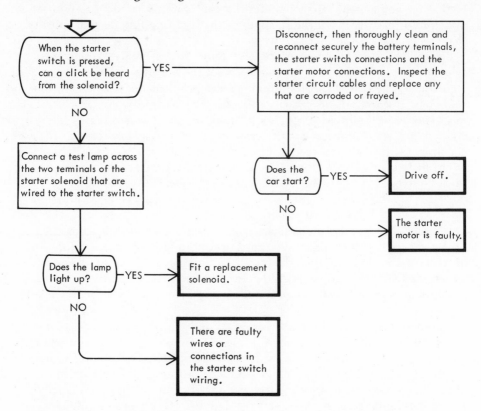

The use of instructions as well as questions in an algorithm is, however, mainly a matter of convenience. Any question implies a test of some kind, and some tests require an action on the part of the tester as well as observation. Suppose, for example, a question in an algorithm were to read 'Is the patient's temperature normal?' Are we seriously to maintain that if the question were to read instead: 'Take the patient's temperature'; 'Is his temperature normal?', the presence of the instruction would be enough to cause us to re-classify the logical tree as an algorithm? If so, this is not a definition which is likely to be of any use to us in the practical writing of algorithms.

In practice the variety of forms possible within the concept of 'algorithm' is far wider than such distinctions would suggest. One of the earliest variants is the List Structure, which, as its name implies, is a re-structuring in list form of the content of the flow chart type of algorithm.

The Rule 5 algorithm above, for example, would appear in list structure form like this:

1. Are you a man?	YES:	Go to 2
	NO:	Go to 3
2. Are you under 25?	YES:	Subscription £2.50
	NO:	Subscription £5
3. Are you under 21 or married?	YES:	Subscription £2
	NO:	Subscription £4

A list structure is certainly easier to print than the flow chart type of algorithm, and may be more suitable for some users. There is, however, no reason why the questions in an algorithm must be connected by lines or must progress down a page. It is, for example, a relatively simple matter to re-arrange an algorithm so as to accommodate it in a wheel or slide device, like this:

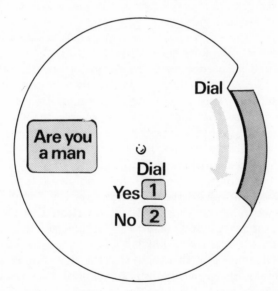

Logically such a device can be identical to the flow chart algorithm and the list structure. Practically it may suit certain users and certain situations better.

Another variant, one that is often used in the analysis of other kinds of algorithm, is the Decision Logic Table. Rule 5, in decision logic table form, would appear like this:

Are you a man?	YES	YES	NO	NO
Are you under 25?	YES	NO	—	—
Are you under 21 or married?	—	—	NO	YES
Subscription £5		✓		
Subscription £2.50	✓			
Subscription £4			✓	
Subscription £2				✓

NOTE: *the dash indicates that the answer may be either YES or NO.*
An explanation of decision logic tables is given in Chapter 3.

So far all the variants of the algorithm we have discussed are identical in logical structure; only the form is changed. There are, however, logical variants too, which are needed for certain kinds of *problem* rather than certain kinds of *user*. We have referred in passing to one of these already: a problem in which one particular case occurs far more frequently than all others. If the algorithm does not take account of this, the user will soon become impatient at having to work through the same series of cases, none of which applies, almost every time he uses the algorithm. We shall deal with frequency-based algorithms in Chapter 5.

A rather similar problem can occur if, for example, an algorithm is designed to locate faults on a large machine or system. Many points on the machine may have to be checked, and if the algorithm is not adjusted accordingly a great deal of effort may be spent hurrying from one point to another and back again, as the technician follows the tests prescribed by the algorithm. The reason for this is not that the algorithm is 'too logical', as is sometimes said, but rather that it is *only* logical; we shall describe in a later chapter ways in which the logic can be retained, while the ergonomic objections are removed.

Finally, the problem itself may be such that a 'normal' algorithm may be less suitable than a means whereby the reader is helped to perform certain mathematical operations. Again we shall analyse this problem later; for the moment it is sufficient to note that the diagram opposite (part of a sequence on rate rebates) is suitable for certain kinds of problem, and that it is an algorithm. It is known as a Guided Calculation.

RECKONABLE RATES

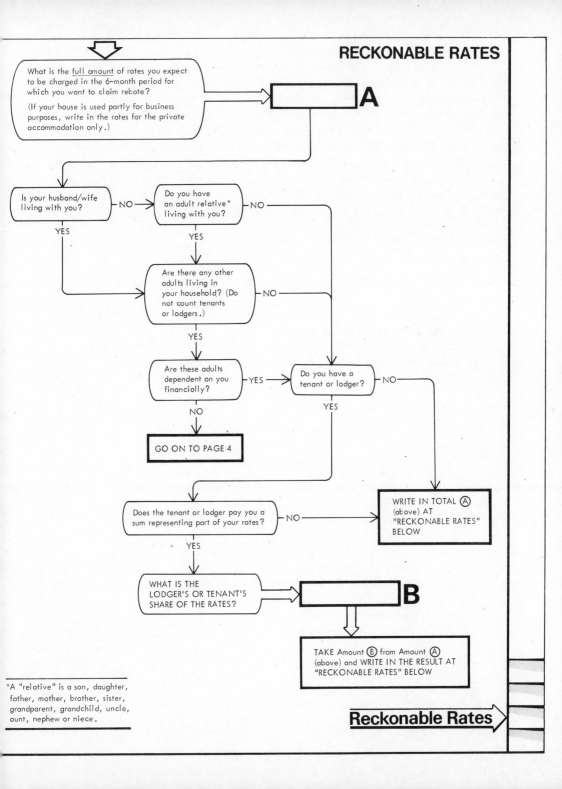

What is the <u>full amount</u> of rates you expect to be charged in the 6-month period for which you want to claim rebate?

(If your house is used partly for business purposes, write in the rates for the private accommodation only.)

A

Is your husband/wife living with you?

Do you have an adult relative* living with you?

Are there any other adults living in your household? (Do not count tenants or lodgers.)

Are these adults dependent on you financially?

Do you have a tenant or lodger?

GO ON TO PAGE 4

Does the tenant or lodger pay you a sum representing part of your rates?

WRITE IN TOTAL Ⓐ (above) AT "RECKONABLE RATES" BELOW

WHAT IS THE LODGER'S OR TENANT'S SHARE OF THE RATES?

B

TAKE Amount Ⓑ from Amount Ⓐ (above) and WRITE IN THE RESULT AT "RECKONABLE RATES" BELOW

*A "relative" is a son, daughter, father, mother, brother, sister, grandparent, grandchild, uncle, aunt, nephew or niece.

Reckonable Rates

Here then is a wide variety of forms into which the same logical structure can be fitted, together with some variants, for special purposes, of the logical structure itself. All are algorithms.

In a developing subject, any definition we propose must itself allow for development. The more rigid it is, the sooner it will be forgotten. At the same time, it must indicate, if possible, the essential quality of the thing described, i.e. the direction in which the most useful development is likely to take place. We therefore propose the following, more as a guide than as a definition: *An algorithm is a means of reaching a decision by considering only those factors which are relevant to that particular decision.*

REFERENCES

Markov, A. A. *Theory of algorithms.* Washington, National Science Foundation, 1961

Wason, P. C. *Psychological aspects of negation.* Communication Research Centre, University College, London, 1962

Lewis, B. N. and Woolfenden, P. J. *Algorithms and logical trees: a self-instructional course. Algorithms Press,* 1969

2 MAKING AN ALGORITHM (1)

How does one set about the job of designing an algorithm? In Chapter 1 we described how one very simple algorithm was built up from a written text which contained all the necessary information. In practice it is by no means common to find a complete text such as this. In many cases there is no text at all, and all the information must be discovered by questioning people who know the work or perhaps by taking the actual machine apart oneself.

Let us imagine for a moment that a training consultant has been given the job of writing a fault-finding algorithm for an electric typewriter. He is questioning an expert and his analysis has just reached the point of considering what could prevent the main power shaft from turning.

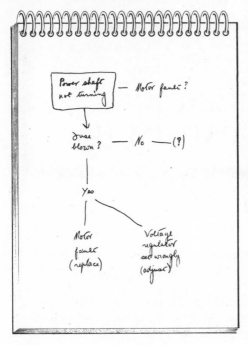

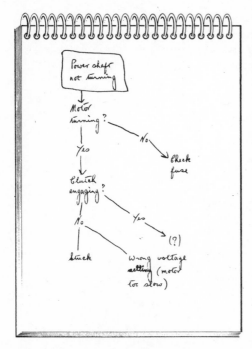

EXPERT: You mean a burnt out motor or something like that?

CONSULTANT: Yes, that could be one reason for the shaft not turning. What other reasons could there be?

EXP: Well, the fuse could go.

CON: That would mean that something else was wrong to cause the fuse to go. Faulty wiring or something?

EXP: No. The machine would never get out of the factory like that. If the fuse goes you can bet it's one of two things: either there's a short circuit on the motor, in which case you'll have to replace it, or else the voltage regulator's set wrongly for the local voltage supply.

CON: Fine. But wait a moment. Why would you think of checking the fuse in the first place? All you know so far, remember, is that the shaft isn't turning.

EXP: Ah, but you could easily see if the motor was turning or not – it's at one end of the shaft.

CON: And if it isn't turning you check the fuse?

EXP: That's right.

CON: Supposing the motor *is* turning?

EXP: Well, it's most likely to be the clutch in that case.

CON: What could be wrong with it?

EXP: It could be clogged with grease. There's a centrifugal clutch on these machines and it sometimes sticks.

CON: Does that happen often?

EXP: No. In fact we find it's usually just that the motor isn't turning

fast enough to operate the clutch –
because they've got it on the
wrong setting at the voltage
regulator. You can see if the clutch
is engaging if you look down
beside the motor.

CON: Right. Now suppose the
clutch *is* engaging, but the shaft
still isn't turning?

EXP: Well, the only other thing
between the motor and the power
shaft itself is the cogged belts. They
could be slipping, or worn out,
or broken.

CON: So if they're not operating
properly you do what?

EXP: Fit new ones. Apart from that
the only thing that could stop the
shaft turning is if something has
jammed it – something dropped
into the typewriter perhaps.

CON: Good. Now I see from my
notes that we've covered the
reasons why a fuse could have
blown and so stopped the motor
and everything else. Now supposing
we find that although the motor
isn't turning the fuse is O.K. Is
that possible?

EXP: Oh yes. The switch contacts
fail to close sometimes.

CON: So if the fuse is all right, but
the motor isn't working, it's got to
be the switch contacts.

EXP: Well, I think you should say
that if it's not the switch contacts
you change the motor. The contacts
are adjustable, by the way. You
don't have to change them normally.

CON: Right. Now we've covered
everything that could stop the shaft
turning. What about if it *does* turn,
but you still can't operate. . . .

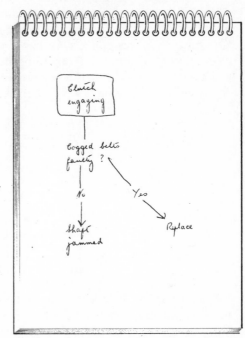

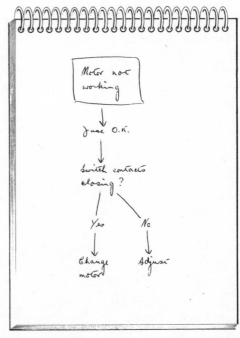

Leaving our consultant to plunge into the next stage of his enquiry, let us see where his questioning has got him so far.

As his final words suggest, he has already isolated the crucial first question in his algorithm. Everything depends on whether or not the power shaft is turning. If it *is* turning, he will have to refer to another algorithm, which we have left him in the process of researching.

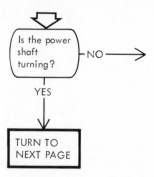

The purpose of his questioning so far is to find out what could prevent the shaft turning. His next job is to turn that information into a diagnostic sequence, progressing in the fewest possible number of steps from symptom to fault. What then is to be the next question on the NO-line? As a general principle, the best question to ask at any point in an algorithm is one which divides the remaining possibilities neatly in half. He notices that the faults described could be divided broadly into electrical and mechanical. Is there one symptom he could ask about in the algorithm which would divide the faults like this? He consults his notes:

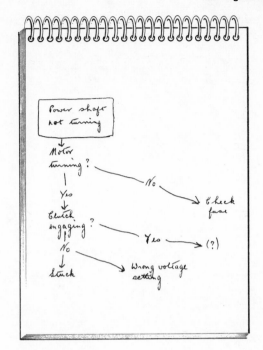

If the motor is turning at all, the enquiry proceeds to the clutch and the cogged belts, which are mechanical. An electrical fault is also mentioned in this sequence (incorrect voltage setting) but he decides to leave this for the moment since its recognition depends on a mechanical symptom (clutch engaging). If the motor is not turning, the enquiry proceeds to the fuse and the switch, which are electrical. This then is his next question:

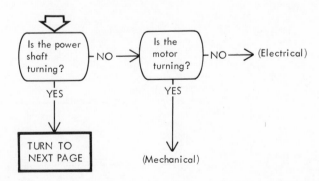

Taking the NO-line first (the motor is not turning), the consultant checks his notes and finds that the next thing to consider is the fuse:

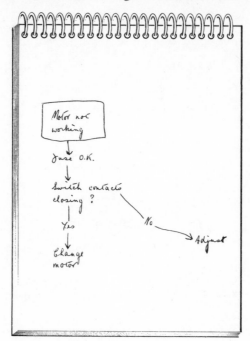

This enables him to fill in four more boxes straight away:

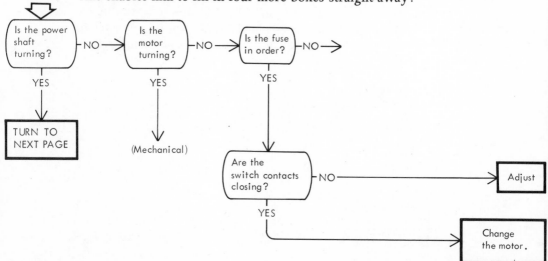

This leaves a loose end on the NO-line. Again he consults his notes:

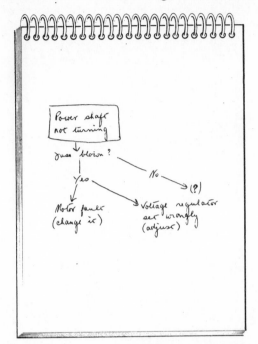

It is clearly much simpler to check that the voltage regulator is set correctly than to test the motor, so he selects this for the next question:

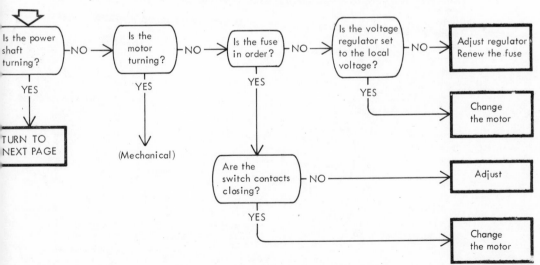

B

This disposes of the electrical side. The first item on the mechanical side mentioned by the expert was the clutch. Here there were two possibilities:

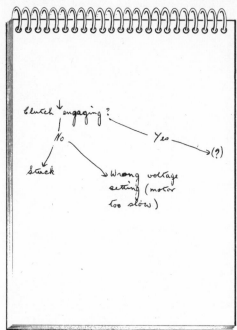

Translated into algorithmic terms, these emerge as:

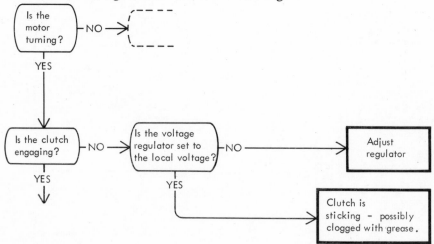

This leaves us with the case in which the clutch engages, but the shaft still does not turn. From the notes it is clear that the next thing to consider is the cogged belts.

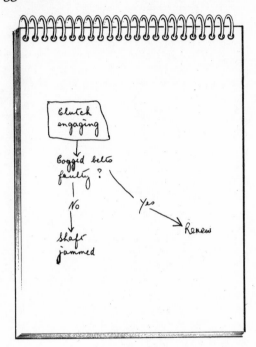

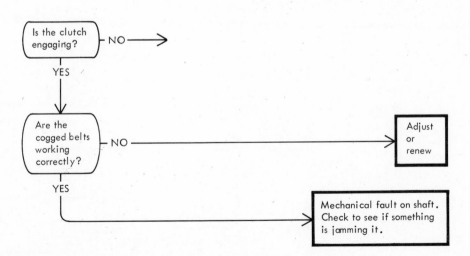

This gives us the complete algorithm:

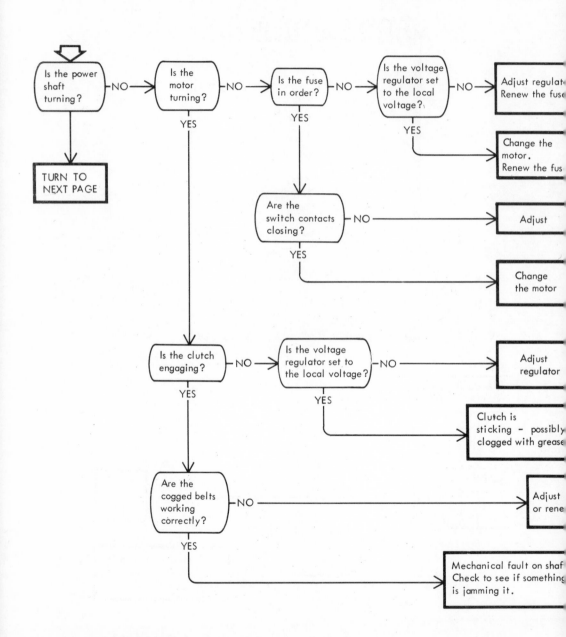

Here then is a simplified though not unrealistic account of how an algorithm comes to be made. Quite a lot of what occurs in practice is omitted: the false starts, the mistakes, the attempts to get concrete information from people who stoutly maintain that what you need to service a machine is not logic but experience, a 'feel' for the job, a sixth sense, or whatever their own speciality happens to be.

However, several points emerge:

1. When getting the information, it is absolutely vital to see that there are no loose ends. If there are, they will emerge in the algorithm as a YES or a NO hanging in mid-air.

2. How to divide up the subject in the form of an algorithm is a problem to which we shall give several answers as this book proceeds. For the moment we shall leave it that as a matter of principle the most effective question at any given point is the one which divides the remaining alternatives neatly in half. (If you have to guess a number, it is much better to ask 'Is it odd or even?' than to ask 'Is it over 10?') If there is not such a question, or if it is impossible to judge which question will divide the alternatives in half, a decision must be made on other grounds, just as our imaginary consultant decided to distinguish between mechanical and electrical faults.

3. In writing algorithms we must always be careful to distinguish between symptoms and faults. A rigid classification into mechanical and electrical *faults* would have prevented us from including an electrical fault (wrong voltage setting, causing the motor to run too slowly) with a mechanical symptom (clutch not engaging) on the mechanical side of the algorithm. When we look at it from the point of view of the logic of the *symptoms*, however, this is the only place we can put it, since there is no other path by which we could diagnose it.

We would like to make one final point before leaving this problem: the finished algorithm is effective only if the user can recognise the parts it mentions – the clutch, the motor, the belts, the switch, etc. For someone who cannot do this, it is no use at all. Like other means of communication, an algorithm must be written for a specific purpose and for a specific audience. This is not, however, to imply that an algorithm could not be written for those who had never seen the inside of an electric typewriter. An algorithm is, after all, a completely practical device: it is a *performance aid*, and if the performance of the reader is improved by modifying it, then it should be modified. There is, for example, no reason why illustration and algorithm should not be combined.

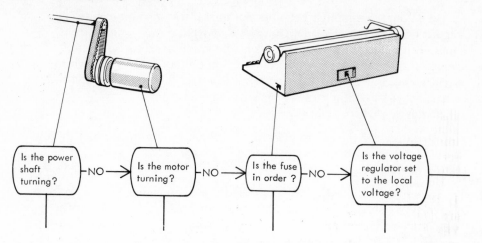

In the following section, which is written in programmed form, you are invited to write an algorithm.

Read each page carefully and decide on the answer to the question at the end of it; then turn to whatever page number is shown there.

Before you start, make sure you have a pencil, a rubber and a sheet of paper on which to write your algorithm.

The following is taken from a leaflet explaining the provisions of the Leasehold Reform Act 1967. This Act gives certain leaseholders the right either to buy their freeholds or to extend their leases.

As you read it, try to decide into what general areas you should divide up your algorithm. For example, we found it helpful to divide up the algorithm on typewriter faults into electrical and mechanical symptoms. Are there, in this case, certain different kinds of qualification for buying a freehold?

WHO QUALIFIES?

1   Have I the right to buy the freehold of my house or extend my lease for 50 years?

YES IF
(a) your lease was originally granted for a term of more than 21 years and
(b) your lease is of the whole house and
(c) your lease is at a low rent. That is, your annual rent is less than two-thirds of the rateable value of your house (as assessed on 23rd March 1965) and
(d) on 23rd March 1965 your house had a rateable value of not more than £400 if it is in the Greater London Area, or £200 anywhere else in the country and
(e) you are occupying the house as your only or main residence and you have lived there either for the whole of the last five years, or for a total of five years during the last ten years.

2   Does the same apply if I am leasing a flat or maisonette?

NO

3   Does it matter if part of the house is sublet?

NO

So long as you occupy part of it as your only or main
residence.

4   I live in a flat above a shop and lease the entire
premises. Do I qualify?

YES

If the building as a whole is a house, and contains your
only or main residence, it does not matter if you use
part of it for a business or other non-residential purpose.

5   I first occupied this house as a weekly tenant and then
bought the lease from my landlord. Can I count my
time as a weekly tenant towards the five years'
occupation which would qualify me?

NO

You may count only the time in which you have been
occupying the house, or part of it, as a leaseholder.

6   I am a recent widow and the lease of the house passed to
me when my late husband died. Do I have to live here
for five years as a leaseholder in my own right to qualify?

NO

Any member of the previous leaseholder's family may
count towards the five years any period in which they
lived with him in the house during the past 10 years.

WHEN YOU ARE READY, TURN TO PAGE 26

In our view, the qualifications for buying a freehold or extending a lease under the Act are of two kinds:

1. Residence qualifications – how long the leaseholder has lived in the house, whether it is his main residence, etc.

2. Qualifications relating to the type of dwelling, its rateable value, etc.

Of course, it is not yet possible to say whether this division is a useful one. This will only emerge after we have made some progress with the algorithm, but it is a starting point. What we have to do now is to decide which kind of qualification to deal with first. Since it does not really matter which we choose, let us start with the residence qualifications.

What are the residence qualifications as expressed in the leaflet?

Make a list.

Then go on to page 27.

Residence qualifications:

1.(e) 'you are occupying the house as your only or main residence and you have lived there either for the whole of the last five years, or for a total of five years during the last ten years.'

5.   'You may count only the time you have been occupying the house, or part of it, as leaseholder.'

6.   'Any member of the previous leaseholder's family may count towards the five years any period in which they lived with him in the house during the past 10 years.'

We have now to decide on the opening question in our algorithm.

Our first task is to think of each of these qualifications as questions in a potential algorithm. The logical interdependence of the ideas must be preserved, but the order in which they are expressed here should be questioned. The final order may, however, in some cases be determined by what is reasonable, rather than what is logically necessary.

Suppose we are to choose between the following as the first question in the algorithm; which do you consider would be the more suitable?

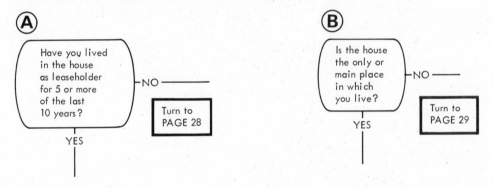

You have chosen

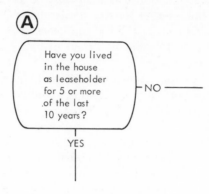

From the strictly logical point of view there can be no objection to this as the first question. Yet would it not be more reasonable to ask first whether the leaseholder is living in the house *at all*, before asking whether he has been living there in the past? After all, many people acquire a lease by buying it or having it left to them in someone's will. The owner of a leasehold house does not, by any means, necessarily live there. For these reasons we would choose question B as the first question in the algorithm.

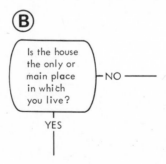

Write this question in the top left-hand corner of your paper.

It is worth noting that we are already in a position to fill in the next box on the NO line.

Decide for yourself what this should be and turn to page 30.

You have chosen

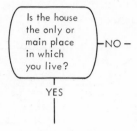

We agree. It is reasonable to ask if the person is living in the house *at all* before asking whether he has been living there during the past ten years.

Write this question in the top left-hand corner of your paper.

It is worth noting that we are already in a position to fill in the next box on the NO line.

Decide for yourself what this should be and turn to page 30.

The next box on the NO line is, of course, the outcome. The regulations clearly state that to be qualified to buy the lease, the house must be the person's only or main residence.

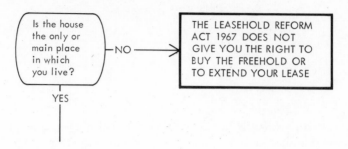

We now have to decide on the next question on the YES line. Let us consider the question we labelled as question A again.

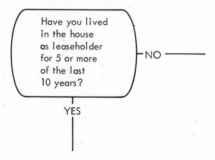

Is this question suitable at all? – as it stands? – or with some modification?

When you have decided, go on to page 31.

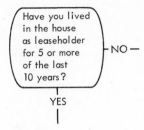

Logically this question is valid. Practically it is too complicated. It is, in fact, two questions rolled into one and this is a probable source of confusion for the person using the algorithm. (Remember that we have not specified precisely who this is to be. We must therefore design the algorithm so that it can be used by the greatest possible number of people.)

Can you see how to divide this question into two? Look back at the residence qualifications on page 24.

Would you choose either of the following as your next question?

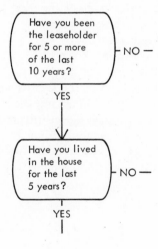

YES: turn to page 32; NO: turn to page 33.

You would choose one of the following as your next question:

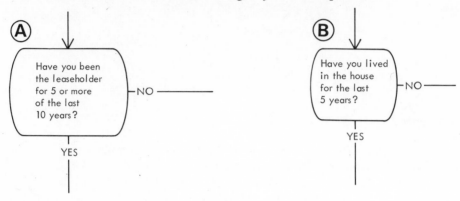

We disagree.

Question A, in our view, causes a break in the sequence. Again it is less a matter of logic than of organisation, but it seems to us more reasonable after a question as to whether the house is the person's only or main residence to go on to ask about residence in the past, before considering the question of ownership of the lease. We certainly have to ask about the lease, but to ask now causes the user to switch his thoughts back and forth unnecessarily.

Question B is totally unnecessary. After asking it, one still has to ask 'Have you lived in your house for 5 or more of the last 10 years?' Since the case of living in the house for 5 of the last 10 *includes* the case of living there for the last 5, there is no point in asking the first question at all.

The fact that this question is taken directly from the original represents a frequently occurring trap for the writer of algorithms. One should always subject the original data to the closest scrutiny. Indeed, it is one of the advantages of algorithms that they highlight deficiencies.

Go on to page 33.

Neither of the following is suitable.

Having asked already about present residence in the house, clearly the next question should concern past residence, and the 5-year qualification. The next question should therefore be:

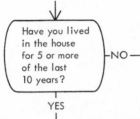

(There is no point in asking 'Have you lived in your house for the last 5 years?', since this is logically included in living there for 5 or more of the last 10.)

To summarise so far, we have written the following questions into the algorithm:

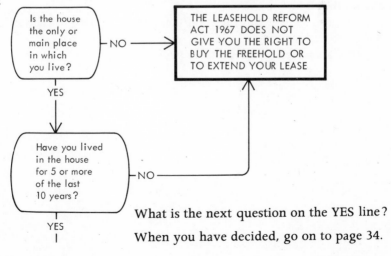

What is the next question on the YES line?

When you have decided, go on to page 34.

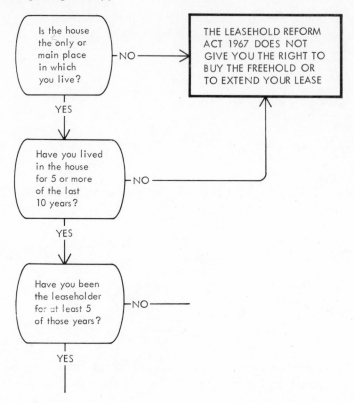

Is the house the only or main place in which you live? —NO→ THE LEASEHOLD REFORM ACT 1967 DOES NOT GIVE YOU THE RIGHT TO BUY THE FREEHOLD OR TO EXTEND YOUR LEASE

YES

Have you lived in the house for 5 or more of the last 10 years? —NO—

YES

Have you been the leaseholder for at least 5 of those years? —NO—

YES

Why have we not continued the third NO line above to the outcome box on the right?

If you can answer this question you should be able to suggest the next stage in the algorithm.

Think about this; then go on to page 35.

We have not continued the line to the outcome box because the questions shown so far do not exhaust the possibilities.

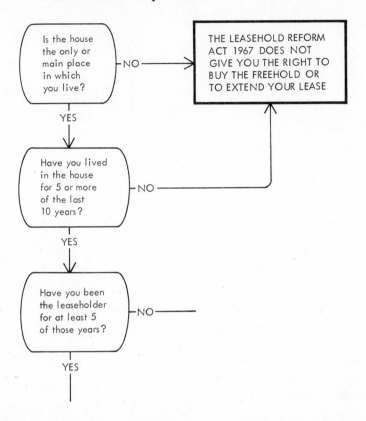

We still have to consider the case of the leaseholder who is not qualified on his own residence as leaseholder, but has lived there as a member of the previous leaseholder's family:

> 'Any member of the previous leaseholder's family may count towards the five years any period in which they lived with him in the house during the past 10 years.'

All this is too much to put into one box if our algorithm is to be useful to many people. There are several possible ways of dividing up this statement.

When you have decided how you would like to do it, turn to page 36.

The problem is to re-shape this requirement of the Act into a series of clear questions and instructions. As often happens in the writing of algorithms, the chief difficulty is to frame the first question in the series. Our solution is as follows:

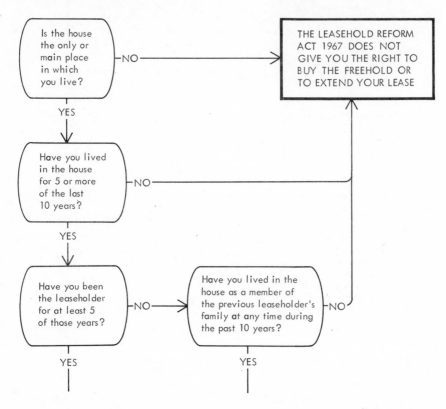

Now fill in the next two boxes on the YES line after the new question. These will contain first an instruction to add the qualifying times together, and then a question as to whether the result is sufficient for qualification under the Act.

Go on to page 37.

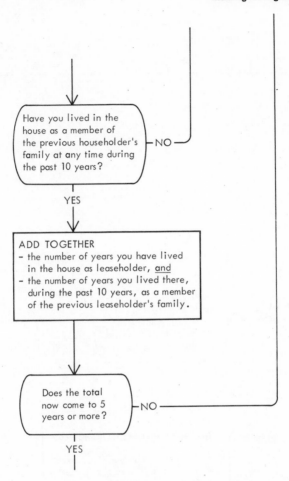

Once again we can continue the NO line to the negative outcome; if the total does not come to five years or more, the leaseholder is not qualified under the Act.

This brings us to the end of the qualifications concerning the leaseholder's residence in the house, and we can pass on to the qualifications concerning the house itself.

The complete algorithm so far is given on page 38.

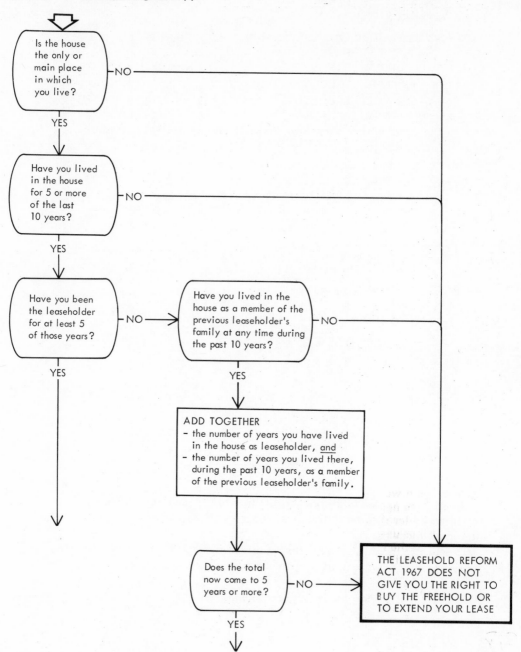

Notice that we have dropped the outcome box so that the whole movement of the algorithm is from top to bottom and left to right. This is not logically necessary, but seems to make for ease of use.

We now pass to the qualifications concerning the house itself.

The first three qualifications concerning the house itself and the lease on it are simple. They are expressed in items 1(a), (b) and (c), with items 2, 3 and 4 as explanatory notes to them. There is no particular reason to place them in any other order, so we can simply write them in as questions.

NOTE: *Item 1(a) refers to the term for which the lease was originally granted. This refers to the length of the original lease, which may, for example have been for 99 years. Since that time the lease may have been bought and sold several times, the price varying according to the number of years remaining on the lease. In this case it does not matter when the lease was taken over provided the original term was more than 21 years.*

Write in the question boxes and any appropriate outcome for items 1(a), (b) and (c).

When you have decided on your first three questions, go on to page 41.

WHO QUALIFIES?

1   Have I the right to buy the freehold of my house
    or extend my lease for 50 years?

    YES IF

    (a)   your lease was originally granted for a term of more
    than 21 years and
    (b)   your lease is of the whole house and
    (c)   your lease is at a low rent. That is, your annual rent
    is less than two-thirds of the rateable value of your house
    (as assessed on 23rd March 1965) and
    (d)   on 23rd March 1965 your house had a rateable value
    of not more than £400 if it is in the Greater London Area,
    or £200 anywhere else in the country and
    (e)   you are occupying the house as your only or main
    residence and you have lived there either for the
    whole of the last five years, or for a total of five years
    during the last ten years.

2   Does the same apply if I am leasing a flat or maisonette?

    NO

3   Does it matter if part of the house is sublet?

    NO   So long as you occupy part of it as your only or
          main residence.

4   I live in a flat above a shop and lease the entire premises.
    Do I qualify?

    YES   If the building as a whole is a house, and contains your
           only or main residence, it does not matter if you use
           part of it for a business or other non-residential purpose.

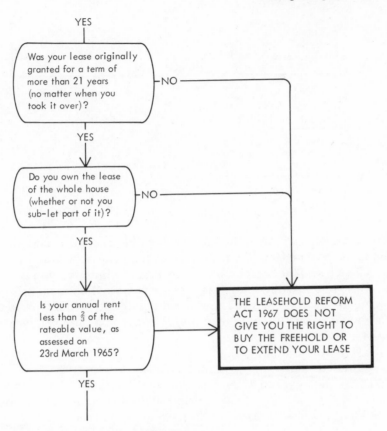

YES

Was your lease originally granted for a term of more than 21 years (no matter when you took it over)? ──NO──

YES

Do you own the lease of the whole house (whether or not you sub-let part of it)? ──NO──

YES

Is your annual rent less than $\frac{2}{3}$ of the rateable value, as assessed on 23rd March 1965? ──→

THE LEASEHOLD REFORM ACT 1967 DOES NOT GIVE YOU THE RIGHT TO BUY THE FREEHOLD OR TO EXTEND YOUR LEASE

YES

Whether or not you add the phrases in brackets in the first two boxes depends on the kind of user you have in mind. Logically they are not necessary, but they may help. This leaves us with item 1(d):

     YES   If . . . (d) on 23 March 1965 your house had a rateable value of not more than £400 if it is in the Great London Area, or £200 anywhere else in the country. . . .

What is the key question for this item?

Go on to page 42.

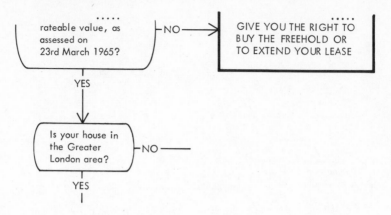

Now add the questions on the YES and NO lines and write in the outcomes. NOTE: You may find it difficult at this stage to link up the negative outcome (**The leasehold reform act does not give you the right . . .**) above. At this stage it is best not to try to link them up. Just write in new outcomes and consider the style and layout separately.

When you have written in the remaining questions and outcomes, turn to page 43.

TO REMIND YOU
YES   If . . . (d) on 23 March 1965 your house had a rateable value of not more than £400 if it is in the Greater London Area, or £200 anywhere else in the country. . . .

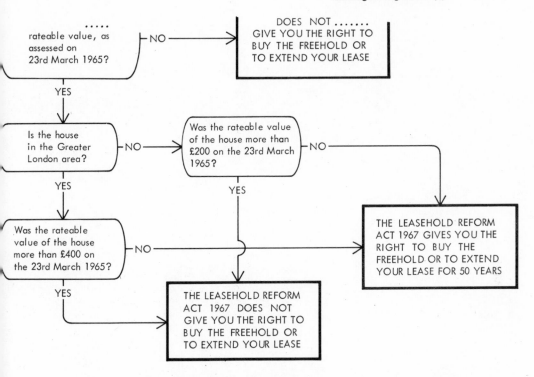

You may have decided to place the YES and NO differently on one of the boxes, so as to avoid the awkward cross-over.

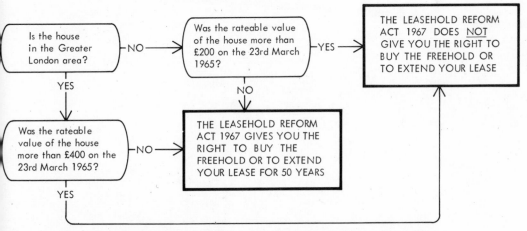

There are several other possible ways of tidying up the algorithm.
You will find the complete algorithm on page 44.

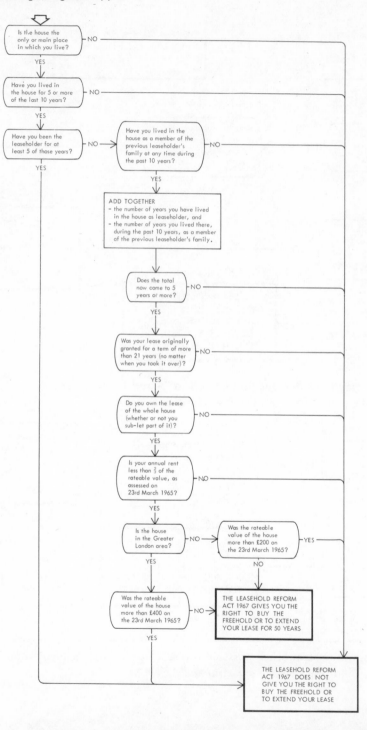

The example you have just worked through should have shown you something of how to put an algorithm together once you have determined what kind of questions to ask. What it could not show you to any great extent is how to work out the questions in the first place.

For this we recommend practice, and for those readers who enjoy puzzles we have constructed the not-too-serious problem below.

The only advice we would offer is that you take special care as you work out your first question.

'By tomorrow', said Marshal Bleu, 'the enemy will be at the river. As I see it, much depends on whether or not he tries to cross. If he does, and we can't destroy the bridge . . .'

'We stand firm!' shouted Marshal Rouge.

'Oh, agreed, provided, of course, the Prussians haven't arrived. If they have, I recommend a withdrawal.'

'You are assuming, I take it, that the enemy has a fairly large force?'

'Yes, I'd set the limit at around 30,000. More than that and I think we should withdraw, in the circumstances we're discussing.'

'But even if he has over 30,000, and even if he tries to cross the river, provided we can blow up the bridge we have no problem. We stand firm.'

Marshal Jaune rattled his sabre. 'In my opinion, if he has less than 30,000 he won't try to cross at all. He knows he wouldn't stand a chance. He'll try to race us to the frontier instead.'

'If he does,' said Marshal Bleu, 'I recommend a withdrawal. If he doesn't, we just stand firm.'

'Ah, but wait,' objected Marshal Rouge. 'Don't forget that even if he has more than 30,000 he still may not try to cross the river. He may think we have a larger force than we have.'

'True,' agreed Marshal Bleu. 'And if he doesn't try to cross we can proceed with our original plan and attack the Prussians.'

The Emperor stood up. 'Gentlemen, I agree with everything you have said. Since we have decided the question of what to do tomorrow I suggest you all get some sleep. Report to my tent at six. Good night.'

'That's the trouble with working for a genius,' sighed Marshal Bleu. 'You need an algorithm to find out what he wants.'

The answer is given on page 46.

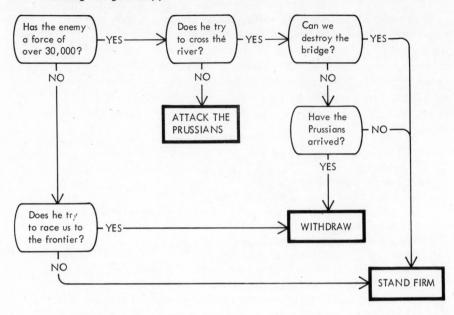

3 MAKING AN ALGORITHM (2): FORMAL METHODS

It is tempting to wonder, as we set about the task of designing algorithms, whether there could not possibly be an algorithm to tell us how to write algorithms. Perhaps that would be asking too much, but surely so logical a structure ought to have some characteristics out of which one could make a system of some kind, which would make the job of writing algorithms quicker and easier.

Thoughts such as these have prompted efforts to construct such a system. Scattered throughout the literature of the subject we find references to 'valuable heuristics', 'powerful techniques' and the like, which have been developed to help in the writing of algorithms. So far as we are aware all such methods involve, in one way or another, the manipulation of questions and outcomes using a decision logic table, such as we referred to as one of many possible types of algorithm in Chapter 1.

Whether such methods as yet represent a practical way of writing algorithms is open to question. There is, however, no doubt that they work, and that they may be capable of further development. Indeed, one of the reasons why we wish to discuss them here is that we will be using a modified form of decision logic table for a special purpose in Chapter 6.

Let us first see exactly what happens when a piece of writing is transformed into a decision logic table and the table in turn is transformed into a flow chart algorithm. We shall use the 'Rule 15' example again, because it is a simple, yet fairly typical, algorithm.

Here, once again, is the original formulation of the rule:

RULE 5: A man pays an annual subscription of £5, unless he is under 25 years of age, when it is reduced by half. A woman pays a subscription of £4, unless she is under 21 years of age, when it is reduced by half, except that when she is married she may pay the reduced subscription irrespective of age.

The first step is to identify the questions:

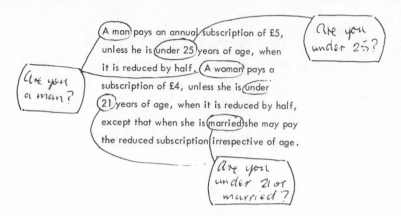

Already we can start to draw up the decision logic table. We know that there are three questions in the algorithm, and that each of these can be answered with a YES or a NO (we do not concern ourselves at this stage with the relationship of the answers to each other). Three questions with two possible answers (YES or NO) to each gives us eight possible combinations (2^3):

Are you a man?	YES	YES	YES	YES	NO	NO	NO	NO
Are you under 25?	YES	YES	NO	NO	NO	NO	YES	YES
Are you under 21 or married?	YES	NO	NO	YES	NO	YES	YES	NO

We now have to isolate the outcomes:

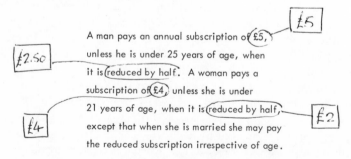

We list the outcomes below the questions and work out from the text which outcome results from each combination of YESs and NOs. There is unfortunately no short cut to this; we just have to look at the text and think hard.

Are you a man?	YES	YES	YES	YES	NO	NO	NO	NO
Are you under 25?	YES	YES	NO	NO	NO	NO	YES	YES
Are you under 21 or married?	YES	NO	NO	YES	NO	YES	YES	NO
Subscription £5			✓	✓				
Subscription £2.50	✓	✓						
Subscription £4					✓			✓
Subscription £2						✓	✓	

From this stage on, however, the construction of the flow-chart algorithm is practically automatic.

The next step is to *identify pairs* of columns, that is to say *columns which lead to the same outcome and are alike in every respect except that one question is answered differently.* The first two columns in the table are an example of this. They both lead to the same outcome (Subscription £2·50) and both have YES opposite the first and second questions, but their answer to the third question is different. *This can only mean that the answer to the third question is irrelevant,* i.e. whichever way it is answered, provided the other two questions are answered by YES, the outcome is the same: Subscription £2·50. We can therefore cross out the irrelevant YES and NO, which leaves us with two identical columns; so we can cross one of these out too.

Pairs: columns which lead to the same outcome and are alike in answers to all questions but one.

Delete the one answer that is different.

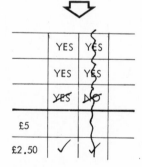

This leaves two identical columns; so we delete one of them.

There are three more pairs in the table and we can deal with these in the same way:

Are you a man?	YES	YES	YES	YES	NO	NO	NO	NO
Are you under 25?	YES	YES	NO	NO	NO	NO	YES	YES
Are you under 21 or married?	YES	NO	NO	YES	NO	YES	YES	NO
Subscription £5			✓					
Subscription £2.50	✓							
Subscription £4					✓			✓
Subscription £2							✓	✓

This effectively reduces the table to four columns only:

Are you a man?	YES	YES	NO	NO
Are you under 25?	YES	NO	–	–
Are you under 21 or married?	–	–	NO	YES
Subscription £5		✓		
Subscription £2.50	✓			
Subscription £4			✓	
Subscription £2				✓

There are no more pairs left in the table and we are now ready to transform the decision logic table into a flow-chart algorithm. Notice that when this stage is reached there is only one complete row of YESs and NOs running across the table. This fact gives us an important clue as to how to start off our flow-chart version: in any algorithm there is only one question that is *always* answered, and that is the first.

In this case the question is:

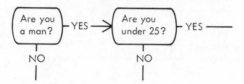

Referring to the decision logic table again, we see that if this first question is answered with a YES the only other question we have to ask is:

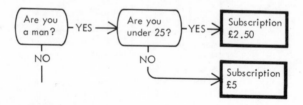

From the answers to this second question, we can pick out the outcomes from the decision logic table without any difficulty: £2·50 follows a YES on the second question and £5 a NO.

The NO-line from the first question box can also be constructed quite automatically from the decision logic table. Those columns which show a NO to the first question contain no reply to the second question at all, so we ignore them. They contain replies to the third question only: 'Are you under 21 or married?'. Here the YES and NO discriminate between the £4 and the £2 subscription. Thus we write them in and arrive at the complete flow-chart algorithm:

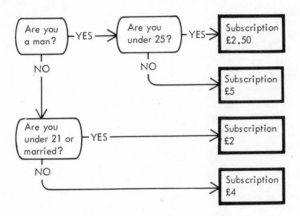

On the following page is a flow chart showing the complete process of constructing an algorithm using a decision logic table.

Identify the questions.

⬇

Write the questions into
the decision logic table.

⬇

Identify the outcomes.

⬇

Write the outcomes into
the decision logic table.

⬇

Write in all possible combinations of
YES and NO for that number of questions. ①

⬇

Work out the outcome for each column
by inspecting the original data.

⬇

Identify pairs (i.e. columns which lead to
the same outcome and are alike in every
respect except that one question is
answered differently) and delete
duplicate columns. ②

⬇

Search for pairs again
in the revised table.

⬇

Are there any? ⟩— YES ——

NO
↓

Begin the flow-chart algorithm by
writing out the question opposite the
only complete row. This is the first
question in the algorithm.

⬇

Find the row which has a reply under
each YES to the first question. This is
the next question on the YES line from
the first question. Write it in.

⬇

Continue in the same way for the rest of
the YES side from the first question.
Continue in the same way for the NO line.

⬇

When all questions are exhausted, write
in the outcomes after the last question
in each line, as indicated by the table.

For example

Question 1	Y	Y	Y	Y	N	N
Question 2	–	–	Y	N	–	–
Question 3	N	N	Y	Y	–	–
Question 4	Y	N	–	–	Y	Y

Here the next question would be
Question 3.

NOTES
1. 'Write in all possible combinations . . .'
The problem is to make sure that one has actually thought of all possible combinations. There should be 2^N columns in the table, where N is the number of questions listed. Thus in the Rule 5 example, we have three questions; so we have $2^3 (= 8)$ columns in the decision logic table. If any reader is dismayed at the practical consequences of this calculation, may we ask him for the moment to continue reading? We shall return to this point later.

2. 'Identify pairs . . .'
The justification for identifying and weeding out pairs is sometimes expressed in terms of Boolean algebra. There is, of course, no need to go to such lengths when actually using a decision logic table to analyse a set of facts and to write a flow-chart algorithm. We present it here for the sake of completeness and in the hope that it may make the point clearer. The method is as follows:

Let us represent the questions in the decision logic table by the letters A, B and C

Question A	YES	YES
Question B	YES	YES
Question C	YES	NO
Outcome	X	X

This is the case we looked at before; the two columns have the same outcome (X) and are identical except in their reply to Question C. In Boolean algebraic form this section of the table would be represented as follows:

$$X = ABC + AB\bar{C}$$

Explanation: 'A' in the expression means that Question A is answered by YES. If it were answered by NO we should put $\bar{A}$, which means 'not A'.

Thus $\bar{C}$ (not C) in the expression records the fact that Question C has a NO opposite to it in the second column.

ABC means 'A and B and C'; + means 'or'.
Thus the whole expression means:

'X equals A and B and C *or* A and B and not C'

This can be simplified as follows:

$$X = ABC + AB\overline{C}$$
$$= AB(C + \overline{C}) \qquad \text{(A and B and C or not C)}$$
$$= AB$$

(That is to say, Question C is irrelevant, as we concluded earlier.)

Similarly we can express the case:

Question A	YES	YES
Question B	YES	YES
Question C	–	NO
Outcome	X	X

Here Question C in the first column can be answered by either YES or NO (i.e. it is irrelevant). The problem is: is the NO opposite Question C in the second column also irrelevant?

The table is converted into the Boolean expression:

$$X = AB + AB\overline{C}$$
$$= AB(C + \overline{C}) + AB\overline{C}$$
$$= ABC + AB\overline{C} + AB\overline{C}$$
$$= ABC + AB\overline{C}$$
$$= AB(C + \overline{C})$$
$$= AB$$

What all this is saying, in mathematical language, is that if a question can be answered either way or not at all, without affecting the result, that question is superfluous to the outcome concerned.

The decision logic table in practice

Having illustrated the method in a simple example, let us see how it works with rather more complex material. We shall use the not-too-serious military example at the end of Chapter 2. Here it is again with the questions identified and marked:

Does he try to cross the river?

Can we destroy the bridge?

Have the Prussians arrived?

Has the enemy a force of more than 30,000

Does he try to race us to the frontier

"By tomorrow', said Marshal Bleu, "the enemy will be at the river. As I see it, much depends on whether or not he tries to cross. If he does, and we can't destroy the bridge ..."
"We stand firm!" shouted Marshal Rouge.
"Oh, agreed, provided, of course, the Prussians haven't arrived. If they have, I recommend a withdrawal."
"You are assuming, I take it, that the enemy has a fairly large force?"
"Yes, I'd set the limit at around 30,000. More than that and I think we should withdraw, in the circumstances we're discussing."
"But even if he has over 30,000, and even if he tries to cross the river, provided we can blow up the bridge we have no problem. We stand firm."
Marshal Jaune rattled his sabre. "In my opinion, if he has less than 30,000 he won't try to cross at all. He knows he wouldn't stand a chance. He'll try to race us to the frontier instead."
"If he does," said Marshal Bleu, "I recommend a withdrawal. If he doesn't, we just stand firm."
"Ah, but wait," objected Marshal Rouge. "Don't forget that even if he has more than 30,000 he still may not try to cross the river. He may think we have a larger force than we have."
"True," agreed Marshal Bleu. "And if he doesn't try to cross we can proceed with our original plan and attack the Prussians."
The Emperor stood up. "Gentlemen, I agree with everything you have said. Since we have decided the question of what to do tomorrow I suggest you all get some sleep. Report to my tent at six. Good night."
"That's the trouble with working for a genius," sighed Marshal Bleu. "You need an algorithm to find out what he wants."

We can write the questions straight into the decision logic table in any order, and plot all possible combinations of YES and NO opposite them. Since there are five questions, the table has 2^5 ($=32$) columns.

Does he try to cross the river?	Y	Y	Y	Y	Y	Y	Y	Y	Y	Y	Y	Y	Y	Y	Y	Y	N	N	N	N	N	N	N	N	N	N	N	N	N	N	N	N
Can we destroy the bridge?	Y	Y	Y	Y	Y	Y	Y	Y	N	N	N	N	N	N	N	N	Y	Y	Y	Y	Y	Y	Y	Y	N	N	N	N	N	N	N	N
Have the Prussians arrived?	Y	Y	Y	Y	N	N	N	N	Y	Y	Y	Y	N	N	N	N	Y	Y	Y	Y	N	N	N	N	Y	Y	Y	Y	N	N	N	N
Has the enemy a force of more than 30,000?	Y	Y	N	N	Y	Y	N	N	Y	Y	N	N	Y	Y	N	N	Y	Y	N	N	Y	Y	N	N	Y	Y	N	N	Y	Y	N	N
Does he try to race us to the frontier?	Y	N	Y	N	Y	N	Y	N	Y	N	Y	N	Y	N	Y	N	Y	N	Y	N	Y	N	Y	N	Y	N	Y	N	Y	N	Y	N

Only three courses of action are suggested by our multi-coloured marshals: attack, stand firm and withdraw. So we can write these in below as the outcomes, and work out from the original text which column goes with which outcome.

Does he try to cross the river?	Y	Y	Y	Y	Y	Y	Y	Y	Y	Y	Y	Y	Y	Y	Y	Y	N	N	N	N	N	N	N	N	N	N	N	N	N	N	N	N
Can we destroy the bridge?	Y	Y	Y	Y	Y	Y	Y	Y	N	N	N	N	N	N	N	N	Y	Y	Y	Y	Y	Y	Y	Y	N	N	N	N	N	N	N	N
Have the Prussians arrived?	Y	Y	Y	Y	N	N	N	N	Y	Y	Y	Y	N	N	N	N	Y	Y	Y	Y	N	N	N	N	Y	Y	Y	Y	N	N	N	N
Has the enemy a force of more than 0,000?	Y	Y	N	N	Y	Y	N	N	Y	Y	N	N	Y	Y	N	N	Y	Y	N	N	Y	Y	N	N	Y	Y	N	N	Y	Y	N	N
Does he try to race us to the frontier?	Y	N	Y	N	Y	N	Y	N	Y	N	Y	N	Y	N	Y	N	Y	N	Y	N	Y	N	Y	N	Y	N	Y	N	Y	N	Y	N
Attack the Prussians																	✓	✓			✓	✓			✓	✓			✓	✓		
Stand firm	✓	✓		✓	✓	✓				✓	✓	✓		✓	✓				✓				✓				✓				✓	
Withdraw			✓				✓	✓	✓				✓			✓				✓				✓				✓				✓

(From this stage on it would be valuable for those who wish to gain experience of decision logic tables to try to work out the flow chart algorithm for themselves, using the outline on page 54 as a guide, and using our illustrations here to check their work.)

The next step is to *identify pairs* in the table, i.e. columns with the same outcome and differing only in the reply to *one* question. We ignore for the moment cases in which more than one question has a different reply.

Does he try to cross the river?	Y	Y	Y	Y	Y	Y	Y	Y	Y	Y	Y	Y	Y	Y	Y	N	N	N	N	N	N	N	N	N	N	N	N	N	N	N	N
Can we destroy the bridge?	Y	Y	Y	Y	Y	Y	Y	Y	N	N	N	N	N	N	N	Y	Y	Y	Y	Y	Y	Y	N	N	N	N	N	N	N	N	N
Have the Prussians arrived?	Y	Y	Y	Y	N	N	N	N	Y	Y	Y	Y	N	N	N	Y	Y	Y	Y	N	N	N	Y	Y	Y	N	N	N			
Has the enemy a force of more than 30,000?	Y	Y	N	N	Y	Y	N	N	Y	Y	N	N	Y	Y	N	Y	Y	N	N	Y	Y	N	Y	Y	N	N	Y	Y	N	N	
Does he try to race us to the frontier?	Y	N	Y	N	Y	N	Y	N	Y	N	Y	N	Y	N	Y	N	Y	N	Y	N	Y	N	Y	N	Y	N	Y	N	Y	N	Y
Attack the Prussians																		✓	✓		✓	✓		✓	✓				✓	✓	
Stand firm	✓	✓		✓	✓	✓	✓		✓		✓	✓	✓							✓					✓					✓	
Withdraw			✓			✓		✓	✓	✓			✓										✓			✓					✓

This gives us a simplified table as follows:

Does he try to cross the river?	Y	Y	Y	Y	Y	Y	Y	Y	N	N	N	N	N	N	N	N
Can we destroy the bridge?	Y	Y	Y	Y	N	N	N	N	Y	Y	Y	Y	N	N	N	N
Have the Prussians arrived?	Y	–	–	N	Y	–	–	N	Y	–	–	N	Y	–	–	N
Has the enemy a force of more than 30,000?	Y	N	N	Y	Y	N	N	Y	Y	N	N	Y	Y	N	N	Y
Does he try to race us to the frontier?	–	Y	N	–	–	Y	N	–	–	Y	N	–	–	Y	N	–
Attack the Prussians								✓		✓	✓					✓
Stand firm	✓		✓	✓		✓	✓			✓					✓	
Withdraw		✓			✓				✓				✓	✓		

Once again we can identify pairs and simplify the table still further. In fact, in this case we shall have to go through the process twice more before we have reduced the table to its final simplification.

In practice, this could all be done on the original table, using coloured pencils, for example. For the sake of clarity we have set it out afresh at each stage.

First simplification

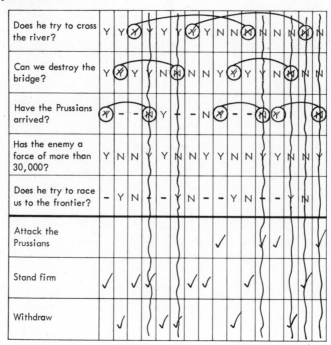

Second simplification

Does he try to cross the river?	Y	Ⓨ	–	Y	–	Y	N	Ⓝ	N
Can we destroy the bridge?	Y	–	Ⓨ	N	Ⓝ	N	Ⓨ	–	Ⓨ
Have the Prussians arrived?	–	–	–	Y	–	N	–		
Has the enemy a force of more than 30,000?	Y	N	N	Y	N	Y	Y	N	Y
Does he try to race us to the frontier?	–	Y	N	–	N	–	–	Y	
Attack the Prussians							✓		✓
Stand firm	✓		✓		✓	✓			
Withdraw		✓		✓					✓

Third simplification

	(1)	(2)	(3)	(4)	(5)	(6)
Does he try to cross the river?	Y	–	–	Y	Y	N
Can we destroy the bridge?	Y	–	–	N	N	–
Have the Prussians arrived?	–	–	–	Y	N	–
Has the enemy a force of more than 30,000?	Y	N	N	Y	Y	Y
Does he try to race us to the frontier?	–	Y	N	–	–	–
Attack the Prussians						✓
Stand firm	✓		✓		✓	
Withdraw		✓		✓		

(1) (2) (3) (4) (5) (6)

We are now ready to write the flow-chart algorithm.

Only one row in this last, fully simplified table is complete. This row, as we found in the last example, must represent the first question in the algorithm, since the first question is the only one which is *always* answered.

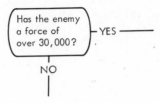

It is useful at this stage to note the numbers of the columns between which the YES and NO discriminate.

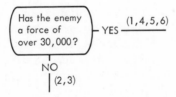

A NO to the first question leaves us with only columns 2 and 3 (and the outcomes associated with them in the table) to consider. From the table it is clear which question will discriminate between these outcomes:

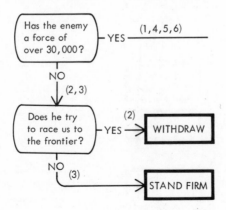

D

We now take the YES line from the first box. To find the next question on this line we search the table for a question which discriminates between columns 1, 4, 5 and 6 (the columns containing a YES to the first question in the algorithm). Only one question has replies in all these columns, so this must be the next question on the YES line from the first box.

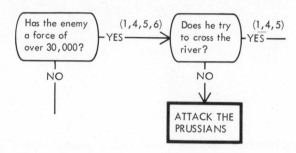

We can write in the outcome for column 6 immediately, since this is now left on its own with no alternative. Again we search the table for a question which has replies in columns 1, 4 and 5. This question is 'Can we destroy the bridge?', and a YES on this accounts for column 1 with its outcome (Stand Firm). Only columns 4 and 5 remain, and the third question in the table discriminates between these.

This gives us an algorithm in rough draft as follows:

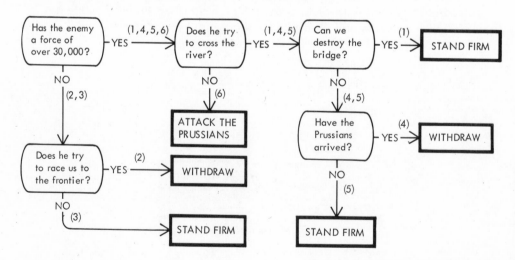

The algorithm as it now stands is logically complete. In practice algorithms are easier to use if they are tidied up a little, particularly by cutting out the repetition of outcome boxes, like this:

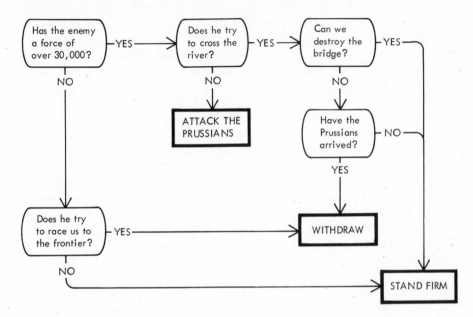

(Note that we have reversed the direction of the YES and NO on one of the boxes.)

Those readers who have worked out this last example both by thinking it through (in Chapter 2) and by using the decision logic table method here, will have realised that there is an important difference in the kind of thinking one has to do with each method, and probably an even greater difference in the time involved.

The decision logic table gives assurance: if it all 'comes out' in the end there can be no question of having overlooked anything, *provided one has identified the questions and outcomes correctly*. On the other hand it may be objected that identifying questions cannot in practice be separated from working out their consequences in combination with each other and with the outcomes; if one works them out without considering this, one runs the risk of misunderstanding them, yet if one understands their consequences in relation to each other one might as well go ahead and write the algorithm straight away, without going to the trouble of working out a decision logic table at all.

Another problem arises with incomplete data. The system of identifying and cancelling out pairs in the table and inspecting it to identify each successive question assumes that the data from which the table was made were complete. In practice this is by no means always the case. Indeed it is one important advantage of using algorithms that they highlight ambiguities and omissions in a set of regulations or whatever the data may be. The omissions may exist in the original text or only in the understanding of the person writing the algorithm; there is no reason why one should not write a flow-chart algorithm to clarify one's ideas and see where the gaps in one's knowledge are. It is not nearly so easy to do this with a decision logic table.

One great disadvantage of the decision logic table method is the amount of time spent in working out possibilities that are not going to be used in the final algorithm. With the five-question algorithm we have examined here, we have had to work out thirty-two possibilities, only to eliminate all but six of them. In practice algorithms which contain as few as five questions are rare; with seven questions one would have to work out 128 possibilities (2^7); with nine questions, 512. By the time the algorithm reaches this size it is obvious that most of the work of constructing the table would be wasted. Every additional question, in fact, doubles the number of columns in the table, and with algorithms of any size more than doubles the number that will eventually have to be deleted. It is tempting to conclude that decision logic tables are so cumbersome as to be practically useless.

We ourselves would not go as far as this. Although the use of decision logic tables in the day-to-day writing of algorithms is limited, they are of value as an aid to understanding the principles on which algorithms are based, and as a means of transforming one kind of algorithm into another, or of modifying it according to constraints imposed by considerations of frequency and ergonomics. For the latter purpose they may be essential, and we shall discuss this use of decision logic tables in Chapter 6.

4 CONTENT AND FORM

Writers on algorithms usually make the point that before an algorithm can be written the problem it is to solve must be well defined: there must be no doubt about what the problem is or what constitutes a solution to it.

The converse is, however, not necessarily also true. We cannot assume that because a problem is well-defined we can just go ahead and make an algorithm to solve it. Still less is it true that if we ask well-defined questions about a problem we can write an algorithm to solve it.

In this chapter we shall consider these cases: a specific type of problem in which both the nature of the problem and the questions we have to ask about it are well defined, yet the problem resists algorithmic treatment; and a further case in which the problem is not easy to define in detail, although we can ask quite specific questions about it, yet it is possible to construct a graphic treatment of an algorithmic kind which goes some way towards solving it.

Both these cases may be considered to be on the fringes of algorithm work proper; indeed many people would claim that one or both of the solutions we are proposing are not strictly speaking algorithms at all. Such problems are, however, central to the working lives of many people, and the solutions we have chosen would probably never have been thought of had it not been for earlier work on algorithms. We discuss them here not only because we believe they are valuable in themselves, but also in order to suggest where the limits of algorithmic treatment lie at present, and to indicate possible directions in which development may take place.

In the first case we are discussing, the original information is from the regulations for Supplementary Pensions and covers the allowance made in respect of rent of accommodation:

'. . . the rent addition will be less than the full amount in any of the following circumstances:

1. if you have a sub-tenant, when the rent paid by that sub-tenant will be deducted;

2. if your rent includes a charge for heating and lighting, when a deduction will be made as these expenses are already allowed for;

3. if anyone else, apart from your wife and any dependent children, is living in your household, when only your share of the rent will be allowed. For example, a married couple paying rent of £1·50 a week with an adult son in their household would be entitled to a rent addition of £1, i.e. two thirds of £1·50.'

At first glance there appears to be no difficulty. There are three conditions which limit payment and three quite precise questions can be asked concerning those conditions:

Do you have
a sub-tenant?

Does your rent
include a charge
for heating and
lighting?

Is anyone else,
apart from your wife
and any dependent
children, living in
your household?

The difficulties occur when we try to set out the algorithm. Which is the first question we must ask? Clearly any of these questions could come first, so let us start with

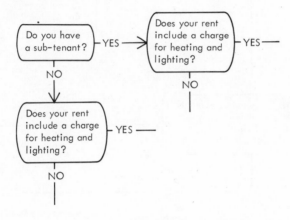

Whether we answer YES or NO to this first question, we shall still have to ask the second question on our list:

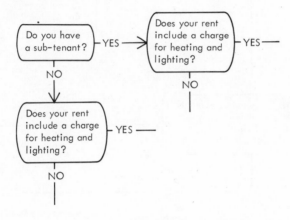

Worse still, the same thing happens with the second and third questions, i.e. none of these questions can be used to exclude any of the others.

By now it is clear that the algorithm is expanding alarmingly. A section written out in full would look like this:

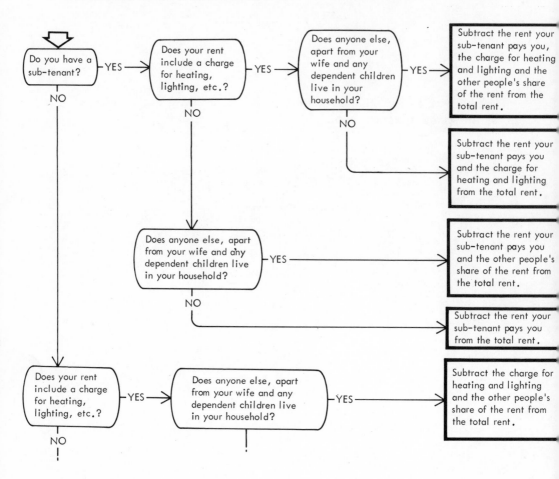

The objections to a layout of this kind are, of course, purely practical. Logically the algorithm works; the user need only answer three questions to arrive at the formula he needs. Such a layout is, however, exceedingly cumbersome, both to construct and to handle. In this particular case, a solution was found by working the problem out in the form of a guided calculation.

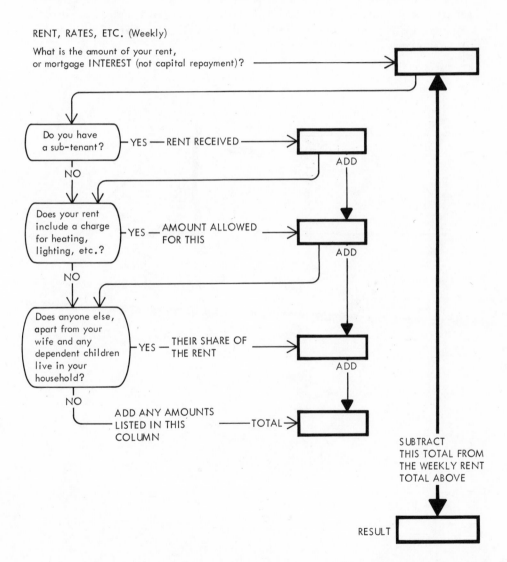

RENT, RATES, ETC. (Weekly)

What is the amount of your rent,
or mortgage INTEREST (not capital repayment)?

Do you have a sub-tenant? — YES — RENT RECEIVED

ADD

NO

Does your rent include a charge for heating, lighting, etc.? — YES — AMOUNT ALLOWED FOR THIS

ADD

NO

Does anyone else, apart from your wife and any dependent children live in your household? — YES — THEIR SHARE OF THE RENT

ADD

NO

ADD ANY AMOUNTS LISTED IN THIS COLUMN — TOTAL

SUBTRACT THIS TOTAL FROM THE WEEKLY RENT TOTAL ABOVE

RESULT

Is a guided calculation really an algorithm? We would answer that it is a means of reaching a decision by considering only those factors which are relevant to that decision; so it is an algorithm.

But then, does it matter whether it is an algorithm or not? A guided calculation is a simple, precise way of proceeding from the facts of the problem to·its solution, so let us use it.

Perhaps it is most helpful to see the algorithm as one example of a more general category of devices which help people avoid mistakes, locate errors, carry out instructions and generally do their work better. The following, for example, is an aid for clerks filling in forms.

One of its pages looks like this, the shaded area representing the form that is slipped under the page so that the part which is to be filled in shows through.

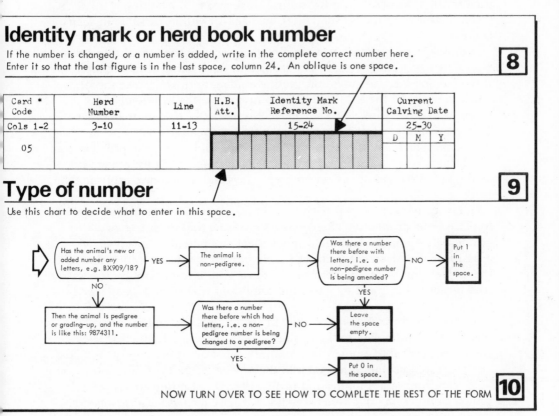

Identity mark or herd book number

If the number is changed, or a number is added, write in the complete correct number here. Enter it so that the last figure is in the last space, column 24. An oblique is one space.

8

Card * Code	Herd Number	Line	H.B. Att.	Identity Mark Reference No.	Current Calving Date		
Cols 1-2	3-10	11-13		15-24	25-30		
05					D	M	Y

Type of number

9

Use this chart to decide what to enter in this space.

NOW TURN OVER TO SEE HOW TO COMPLETE THE REST OF THE FORM **10**

Here we have a device that is obviously akin to an algorithm, but is unlike what most people would readily identify as such. It was designed for a specific purpose, for a specific set of users, for a specific problem, and it enables these users to solve that problem.

There is, in our view, little point in classifying such devices as algorithms or anything else. The point is to make more effective devices of this or any other kind, as aids to communication. It is, after all, the drive for more effective explanation that has led to the development of algorithms themselves.

A second case in which an attempt to use a normal algorithm fails is that in which the objectives cannot be defined closely. This, of course, is well known. The problem is whether simply to say that algorithms cannot be used for that purpose, or to try to find some means of adapting them, or at least to devise something similar that will be of value in such a context.

A situation such as this arose in a banking organisation operating overseas. Potential managers had to be trained in the lending of money to firms. If the lending were too rash, the bank would suffer losses when the firms failed to repay; if the lending were too niggardly, the bank would lose business to other banks. Some managers were better lenders than others. Trainees found great difficulty in grasping the situation at all.

'Conventional' algorithms were clearly impossible here; the problem was too diffuse. What was needed was a means of training the judgement of lending bankers so as to avoid both extremes and to optimise profit for the bank.

In favour of such an approach was the fact that some managers knew how to place loans profitably. There was also the fact that certain tests were customarily applied when the balance sheet of a potential borrower was examined. The results of such tests were never regarded as conclusive in themselves, but were 'interpreted' by those whose responsibility it was to lend or not to lend, and to assess the conditions that should be attached to the loan.

Those managers who were expert in assessing potential borrowers, therefore, were the key to the problem. Only they could say what tests should be applied to a potential borrower, what questions should be asked and what weight should be given to the replies. After a considerable amount of investigation, questioning and testing, a series of rating charts was produced involving questions which the trainee manager had to answer, not with a simple YES or NO but with an assessment expressed as a number.

A	Safety of Loan

1

What is the ratio of Current Assets to Current Liabilities?

	Grades	Mark Grades Below − +
Up to 0.75 : 1	(-8)	☐
0.75 − 1.0 : 1	(-4)	☐
1.0 − 1.5 : 1	(0)	
1.5 − 2.5 : 1	(4)	☐
2.5 − 3.5 : 1	(8)	☐
3.5 or more : 1	(16)	☐

2

Has this Ratio changed over the past 3 years?

Ratio under 2 : 1	Diminished	(-4)	☐
	Stagnating	(-2)	☐
	Increased	(2)	☐
Ratio over 2 : 1	Diminished	(-2)	☐
	Stagnating	(0)	
	Increased	(4)	☐

3

What would be the ratio of Current Assets to Current Liabilities if the loan asked for were granted?

Up to 0.75 : 1	(-4)	☐
0.75 − 1.0 : 1	(0)	
1.0 − 1.5 : 1	(4)	☐
1.5 − 2.0 : 1	(8)	☐
2.0 or more : 1	(16)	☐

4

What is the ratio of Current Assets to Total Liabilities?

Up to 0.5 : 1	(-4)	☐
0.5 − 0.75 : 1	(0)	
0.75 − 1 : 1	(4)	☐
1 − 1.5 : 1	(8)	☐
1.5 − 3 : 1	(12)	☐
3 or more : 1	(16)	☐

5

Was the stated purpose of the loan to increase Fixed Assets?

Fixed Assets which involve significantly more capital outlay, e.g. enlargement of factory, special-purpose buildings, complete modernisation of works.	(-8)	☐ ☐
Fixed Assets which do not involve much further capital outlay, e.g. a simple warehouse, new inexpensive machines which greatly increase efficiency, a new railway siding.	(0)	
	Total	☐ ☐

is (−) greater than (+) ⟶ **Beware of Lending**

is (+) greater than or equal to (−) ⟶ **Go on to Test B**

At first sight it might seem that this merely replaces the one general judgement of the lender's credit-worthiness with a series of smaller judgements and assessments. In fact it was precisely this putting together of many judgements and assessments on details that managers found so difficult. It is a comparatively simple matter to decide what is the ratio of a firm's assets to its liabilities. What is difficult is to assess the importance of this ratio in that particular firm and in that business environment. This is what the rating charts helped the manager to do.

Are charts of this kind algorithms? According to our own definition they are not: they are not *means* of reaching a decision by considering only those factors which are relevant to that decision. They are in fact *aids* to reaching a decision, and their greatest use is in the period of training of each manager, although they are also of value in the initial stages of doing the job. The rating charts cannot be taken by a clerk and used with as much effectiveness as if the manager were doing the job. The manager's judgement is still needed; the purpose of the charts is to guard him from mistakes as he settles into his new job, and to cause him to become accustomed to paying attention to all relevant factors when making his decisions.

One final point: if we examine question 1 on the chart, it is clear that this could be broken down into six separate questions:

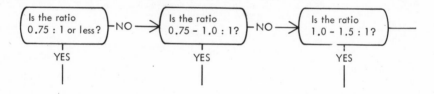

Had this been done, we should have arrived at a 'normal' algorithm, although it would have been a very unwieldy one. To complete the algorithm we should have had to ask all the sub-questions in question 2 *on both YES and NO lines*, and the same would have applied to questions 3, 4 and 5.

Here, then, we have a situation similar to that of our first example in this chapter, except that in this case we are dealing with separate assessments rather than a single calculation broken into parts. The solution in the present case is to grade each assessment, using numerical weightings; this method was applied in matters of pure judgement too.

General assessment of applicant

---

already a customer, how would you grade him compared with other customers?

(Consider the behaviour of his account, the quality of the bills he presents for processing, the amount of business he does with the bank, whether he has honoured his agreements in the past, etc.)

Leave items blank if no information is available

very poor customer			very good customer				−	+
Grade: − 20	− 10	0	+ 10	+ 20			☐	☐

Once again we are crossing a border out of the area of algorithms proper, this time into the field of training. The use (or misuse) of algorithms in training is, however, a subject in itself, and we shall discuss it in a later chapter.

Supplementary note: The guided calculation shown on page 73 has already found a number of applications. It is interesting to speculate what the list structure equivalent of a guided calculation would look like. The following is the nearest practical equivalent we have found. It makes use of a scoring system based on the intervals of binary arithmetic to specify the outcomes. Like the guided calculation it can be printed in far less space than the original algorithm.

RENTS, RATES, ETC. (section of instructions for calculating Supplementary Benefits)

QUESTIONS:

Do you have a sub-tenant?

YES — Score 1; NO — Score 0

Does your rent include a charge for heating, lighting, etc.

YES — Score 2; NO — Score 0

Does anyone else, apart from your wife and any dependent children, live in your household?

YES — Score 4; NO — Score 0

Add up your total score.
Choose the answer below with the same number as your total score.

ANSWERS:

0   The full amount of your rent qualifies for benefit.

1   Subtract the rent your sub-tenant pays you from the total rent. The rest qualifies for benefit.

2   Subtract the charge for heating and lighting from the total rent. The rest qualifies for benefit.

3   Subtract the rent your sub-tenant pays you and the charge for heating and lighting from the total rent. The rest qualifies for benefit.

4   Subtract the other people's share of the rent from the total rent. The rest qualifies for benefit.

5   Subtract the rent your sub-tenant pays you and the other people's share of the rent from the total rent. The rest qualifies for benefit.

6   Subtract the charge for heating and lighting and the other people's share of the rent from the total rent. The rest qualifies for benefit.

7   Subtract the rent your sub-tenant pays you, the charge for heating and lighting, and the other people's share of the rent from the total rent. The rest qualifies for benefit.

5 LOGIC AND ITS MODIFICATIONS – FREQUENCY

Not only the type of information and the type of user can influence the make-up of the algorithm. The way in which the information is collected can have its effect too, and as writers of algorithms we must take care that this is the effect we want.

This is particularly important where frequency is concerned. There is, for example, an obvious difference between the kind of information one gets about a machine from taking it to pieces oneself, and the kind one gets from questioning an expert on likely faults and their detection. Anyone who has no practical experience of servicing a machine and has to rely on his own understanding of its working as he examines it or takes it to pieces will virtually have to assume that all faults are equally likely to occur. A drive belt may break before the switch contacts become dirty; an oil channel may become blocked before a cam wears out; a flexible connector may break before a gasket gives way; he has no way of knowing which will happen first, and his algorithm is therefore arranged in a sequence dictated by the logical structure of the machine alone. The expert, on the other hand, knows that the flexible connector breaks approximately once a fortnight. He also knows that it is an extremely difficult and dirty job to check whether the cam has worn out. In practice he always checks the flexible connector first, because it often goes and is very easy to check, and he leaves the cam until last, even if he suspects it may be worn, because it is easier to check the other faults and there is always a chance that one of them is causing the trouble.

The same writer, therefore, might produce two quite different algorithms on the same machine according to whether he examined the machine himself or asked the expert for information about it. In the first case the algorithm would rely on a simple logical progression, while in the other the logic would be modified by considerations of frequency and ergonomics. In practice, of course, it is best both to examine the machine and to ask the expert. The important thing is to produce an algorithm that is practical and that people are going to use – and it is quite clear that people are not going to use an algorithm that forces them, every time the machine goes wrong, to examine two or three things which are practically always correct before they touch the one thing that usually breaks down.

In this chapter we shall consider the question of frequency and how the frequency with which a fault occurs can be built into an algorithm. To do this we shall look at a case involving weaving looms. The problem here was to train overseers to locate faults quickly, so as to avoid a break in production, and incidentally a loss of bonus for the worker on the loom. The traditional way of training the overseers was for them to work with a skilled man and 'gain experience'. This method proved unduly slow and the degree of training received was very uneven.

An attempt was then made to improve fault location by using algorithms. It was noticed at an early stage that when the experienced men checked for faults they often behaved in an apparently random manner, explaining their behaviour as the result of 'experience'. After a lengthy analysis it emerged that what they were doing was to allow for the fact that some faults occurred more frequently than others. In such a context a set of algorithms that assumed equal frequency of faults would quickly have been discredited and thrown on the scrap heap. It was decided, therefore, to build frequency into the algorithms. The difficulty was that no one could give reliable advice on how often each fault occurred.

The first stage was to examine the looms with an expert and to design algorithms which proceeded in the normal manner from symptom to fault. No attempt was made to allow for frequency at this stage.

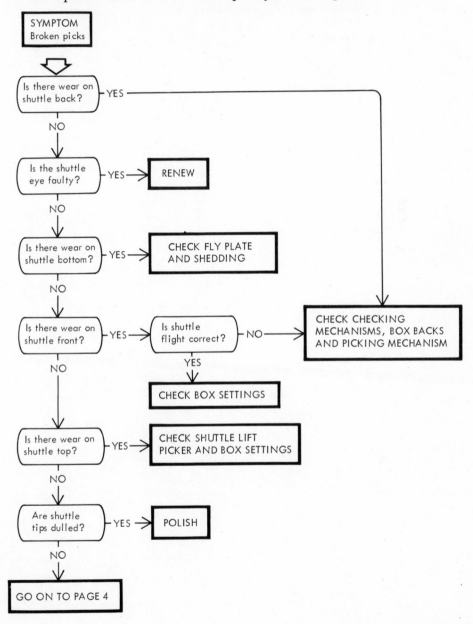

The algorithms were then issued to those responsible for locating and repairing faults. Checking boxes were printed down the right-hand side of each algorithm and each time a fault occurred the person locating it ticked one of these. A typical result is shown below:

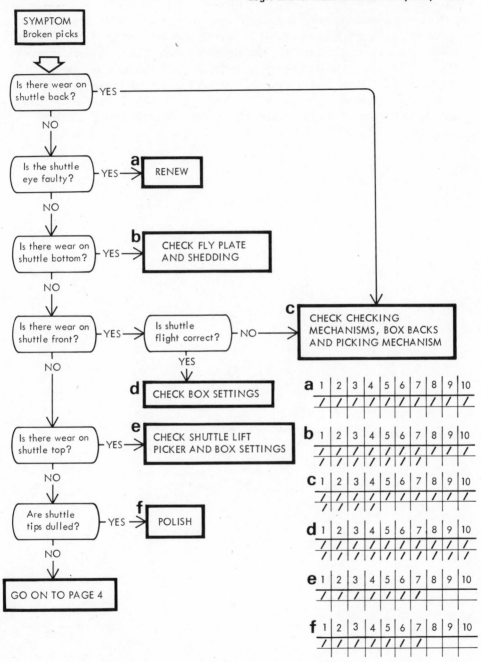

Running down the page, the faults occur in frequencies of 10, 17, 14, 20, 7 and 7. Clearly the first and fourth items on this list are out of place. More often than not, the person using the algorithm would have to work through the first three items unnecessarily in order to reach the most frequently occurring fault. The remedy is simple: items 1 and 4 can be swapped over to give a progression of frequencies as follows: 20, 17, 14, 10, 7 and 7. The algorithm too can easily be adjusted to accommodate this progression of questions:

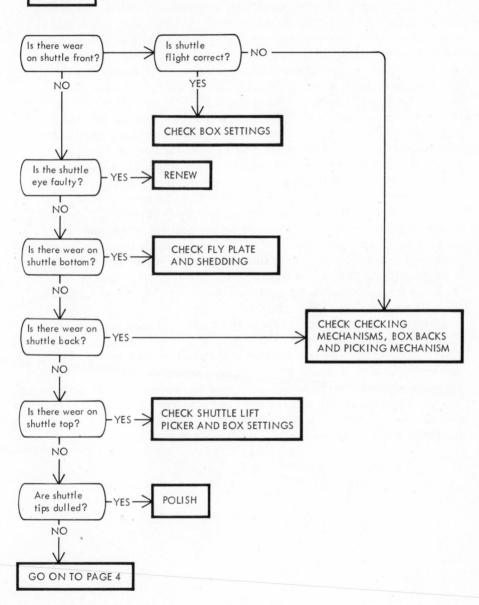

Here, then is a simple example: an algorithm with the limited aim of leading the worker from symptom to fault or to a number of checks which will reveal the fault, all arranged in approximate order of fault frequency. The algorithms work, but, it is reasonable to ask, why stop at this point? Some of the outcomes seem rather indeterminate: CHECK FLY PLATE AND SHEDDING. Which of these is more likely to be wrong?

Unfortunately, the point is sometimes reached in machinery when all that it is possible to say is that a certain selection of faults is possible. This is particularly liable to happen where adjustments are concerned. Of course, with an extensive statistical study it might be possible to discover that, say, given a certain set of symptoms one fault was twice as likely to arise as another. It might indeed be possible to discover the probability of all possible faults arising and to reconstruct the algorithm on this basis.

In practice the law of diminishing returns operates. To progress far in this direction is expensive in time and there is rarely much gain in effectiveness, since few machines are sufficiently consistent in their faults. Besides, if it were really possible to make out a reliable list of possible faults in strict order of probability there would be little incentive to write an algorithm at all; in most cases the list itself would lead one to the fault more quickly. All this, of course, assumes the rather unlikely case that the statistical study is finished before the manufacturer has decided to modify the machine and so invalidate the study.

All things considered, it is cheaper and quicker to make a reasonable allowance for frequency where the conditions of the job require it. We must conclude therefore that the writer of the algorithms for fault finding on looms probably made the best practical decision in leaving his outcomes not quite as precise as one would ideally wish. Whether he was in fact right depends on his aims and whether or not he achieved them, and this in turn can be assessed quite precisely by measuring the improvement (if any) in the performance of the workers who are using the algorithms.

6 LOGIC AND ITS MODIFICATIONS – ERGONOMICS

The simplest way of explaining the ergonomic factor in the design of algorithms is to start with an example. Below is an entirely fictitious algorithm which purports to locate faults in a dish aerial on a tower.

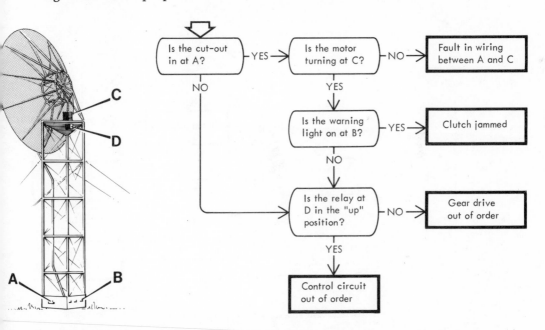

This algorithm, we will assume, expresses the best logical way to find the cause of a malfunction. What happens, then, if we use it to diagnose a fault?

Let us suppose that the aerial fails to turn. We investigate. We find that the cut-out is *in,* so we answer YES to the first question. We are then asked whether the motor at C is turning. So we shin up the tower and find that it *is.* This leads us to: 'Is the warning light on at B?'. So we descend and find out that it is not. Unfortunately, the question this leads to – 'Is the relay at D in the "up" position?' – causes us to shin up the tower once again. Of course, if the algorithm has been correctly designed we can guarantee that we shall find the fault this time, but after two trips up the tower we are not likely to be in a mood to appreciate the logical elegance of the algorithmic system any more. In practice it would be a simple matter to rewrite the algorithm to take this difficulty into account. The obvious thing to do is to check whether the warning light is on at B while we are checking the cut-out at A. As it happens, according to the original algorithm we need only check the warning light when the cut-out is in, i.e. question B appears on the YES line from question A only. This is done in the revised version below, which is known as an *operational* algorithm.

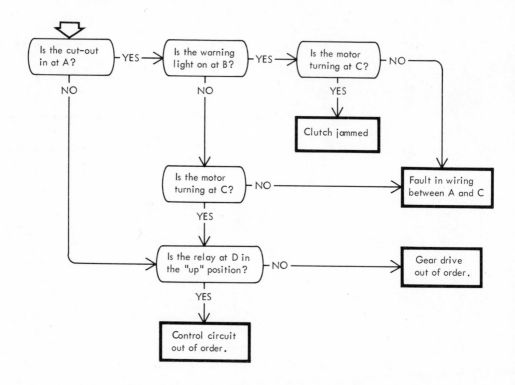

Here the question 'Is the motor turning at C' has been repeated on both YES and NO lines from question B. This is necessary, since the answer to question B in no way influences whether or not we ask about C.

In a simple example such as this there is no difficulty in determining which questions should be repeated and what is the simplest and most economic way of arranging those questions in the operational algorithm. The following example is, however, not nearly so simple. (For convenience we have not written in actual questions. 'A?' means a question about checkpoint A, etc.)

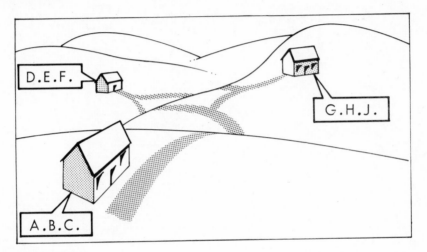

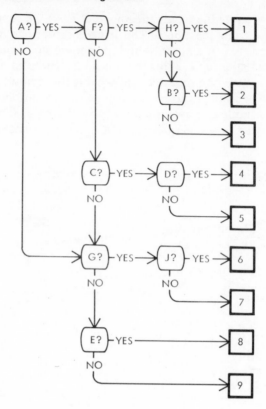

It might seem at first glance that if we write an operational algorithm from this there is a danger of the algorithm expanding to unmanageable size, as we ask each question in each group whatever the answer to the previous questions, just in case we need the information. On closer inspection, however, it emerges that we can exclude some possibilities straight away. If we answer NO to 'A?', for example, we need not ask about 'B', 'C', 'D', 'F' and 'H' at all. If we answer YES to 'A?', any question at all may follow, but if we have asked 'C?' and the answer was YES, we can exclude either 'B?' or 'H?' by asking 'F?'. In view of this, is it better to ask the questions in the order:

	'A?'	'B?'	'C?'
or	'A?'	'C?'	'B?'

or does it not matter?

Even if the answer to this is obvious to the reader, it is certainly no easy task to construct the *simplest possible* operational algorithm by modifying the original according to the ergonomic constraints imposed by the layout of these check-points.

It is worth noting that this task can, however, be done quite mechanically, using a decision logic table. In our previous discussion of decision logic tables in Chapter 3 we came to the conclusion that to use them for the initial drafting of algorithms was likely in most cases to be too cumbersome to be practical. Here, however, the algorithm is already written. All we have to do is to convert it into a decision logic table – a relatively simple process – and to use the table as a means of re-arranging the questions into 'operational' order. This method is particularly useful here since it gives us complete assurance that we have taken all possibilities into account.

First, a decision logic table is drawn up with the questions grouped according to the constraints of the proposed operational algorithm. The answers to the questions are then written in, leaving blanks where no answer occurs in the algorithm.

	1	2	3	4	5	6	7	8	9	
A?	Y	Y	Y	Y	Y	Y|N	Y|N	Y|N	Y|N	First Group
B?		Y	N							
C?				Y	Y	N	N	N	N	
D?				Y	N					Second Group
E?								Y|Y	N|N	
F?	Y	Y	Y	N	N	N	N	N	N	
G?						Y|Y	Y|Y	N|N	N|N	Third Group
H?	Y	N	N							
J?						Y|Y	N|N			
Out-comes	1	2	3	4	5	6	7	8	9	

The last four columns are each divided into two because there are two routes to each of these outcomes.

A draft operational algorithm is then written as follows:
1. Write down the first question:

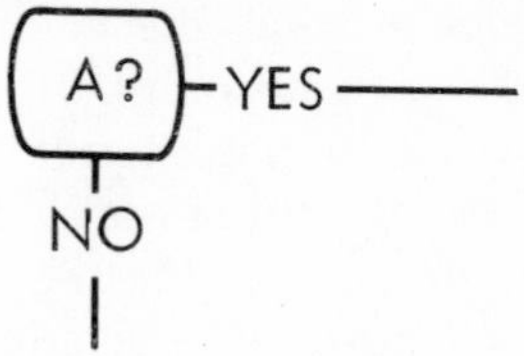

2. Note on the YES and NO lines the possibilities between which this question discriminates:

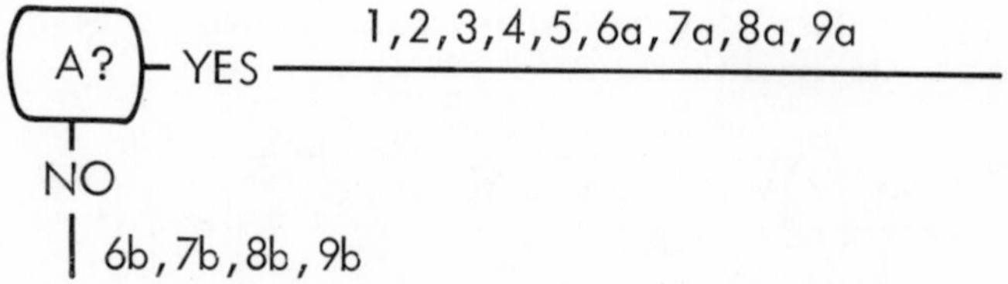

The 'a' and 'b' refer to the divided columns. Remember also that in a decision logic table where no YES or NO appears, the reply must be assumed to be 'either YES or NO'.

3. Select the next question on the YES line *from the same group* as the first question. This must be the question which discriminates between the greatest number of outcomes – i.e. the line with the greatest number of YESs and NOs in it in that group. Here it is question 'C?'.

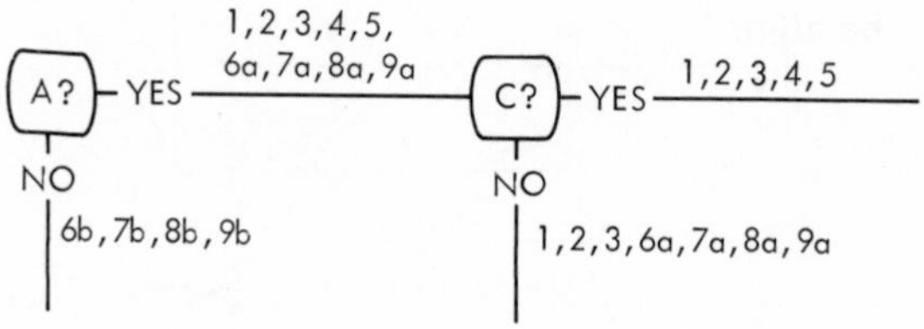

4.   Take the YES and NO lines from the second box and go through the same process. In this case it is necessary to ask question B on both lines, since B? has outcomes in the columns listed there.

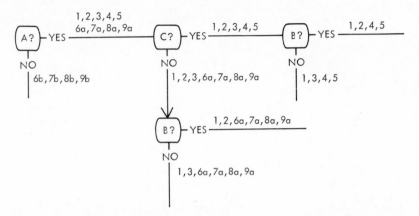

5.   Go through the same process for the NO line from the first box. In this case no further questions in that group need be asked, since no replies appear in the table under the NO answer to the first question.

6.   When the first group is exhausted, go on to the second. Continue until the whole table is exhausted and write in the outcomes.

This gives us a draft operational algorithm as follows:

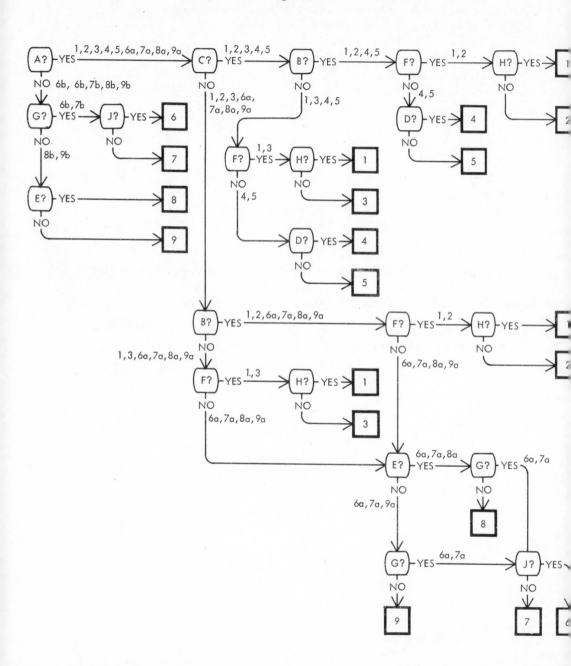

The draft can be tidied up by cutting out repetitions and running lines together wherever this can conveniently be done.

Tidied-up operational algorithm

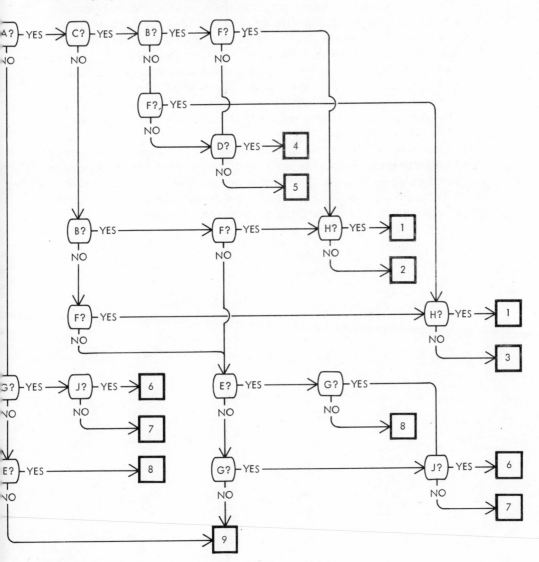

In our formalised version this is as far as we can go. In practice, it would often be possible to simplify the algorithm further. If the real question 'B?', for example, were 'Is the fuel tank empty?' there would be no need to ask any further questions on the YES line at all.

Here then is the full method and justification for making operational algorithms. It is, of course, only rarely that one need go to such lengths as this. Often the ergonomic factor emerges in the form of an isolated obstacle to the progress of the algorithm; having written it the experts point out to us that one of the checks we propose in the middle of the sequence is very difficult to do and is in practice always left until last for this reason. A case such as this presents the algorithm writer with an opportunity to rationalise the work sequence. By applying the principles outlined above, even if he does not need to use the full decision logic table treatment, he may well be able to suggest a more economic sequence of operations than the practitioners themselves are using.

For those who wish to try their hands at the full technique for converting algorithms into their operational equivalents, we conclude this chapter with a further formalised example. This is divided into four parts: the original algorithm and the ergonomic constraints; the decision logic table; the draft operational algorithm; and the finished operational algorithm.

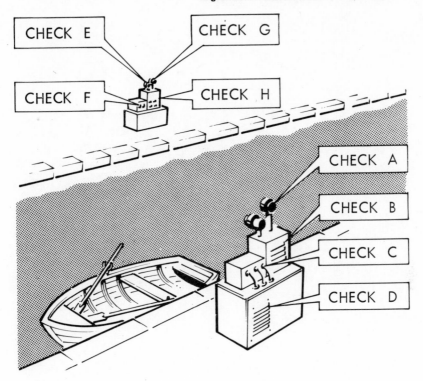

**The
original
algorithm**

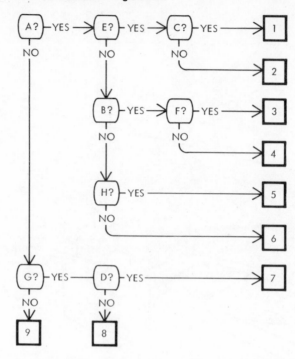

The problem: To re-arrange the algorithm so as to locate all faults
(numbered 1–9) with the fewest possible trips across the river.

The decision logic table

A?	Y	Y	Y	Y	Y	Y	N	N	N	
B?			Y	Y	N	N				
C?	Y	N								GROUP 1
D?							Y	N		
E?	Y	Y	N	N	N	N				
F?			Y	N						
G?							Y	Y	N	GROUP 2
H?					Y	N				
Out-comes	1	2	3	4	5	6	7	8	9	

The draft operational algorithm

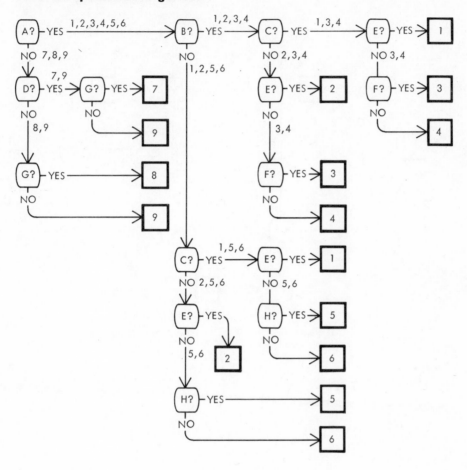

The finished operational algorithm

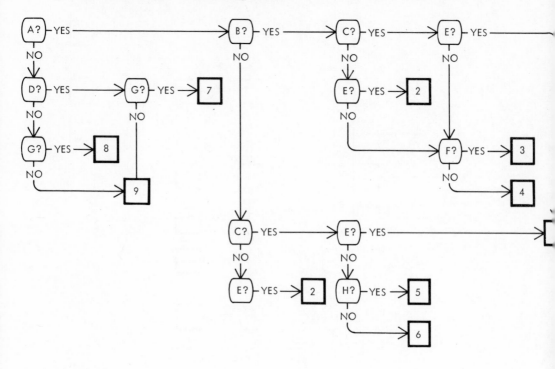

7 WHAT CAN ALGORITHMS BE USED FOR?

'Can you write an algorithm for . . . . . ?' The question, sometimes facetious, sometimes exploratory, often in earnest, is continually asked of the designer of algorithms. The subjects envisaged may vary from how to deal with doorstep salesmen to landing an aeroplane, from diagnosing faults in cakes to choosing a girl friend, and the answer we give may vary accordingly.

Sometimes we feel able to reply with a defninite 'No'; after all, if the subject itself is not based on logical decision making, there is no point in trying to express it logically. Hamlet's 'To be or not to be . . .' may start with a question, but there is no point in trying to make an algorithm out of it. In other cases we are less certain. To quote an example we have already looked at in Chapter 4: is it clear whether or not we can write an algorithm to enable bank managers to assess firms for loans? Could we even design a strictly logical flow chart of the process, and if we could how would we know whether it was complete, and if it were complete would it not be of such size and complexity as would make it extremely laborious to use? We can certainly list the main questions and give guidance as to what weight should be given to each factor. The result, part of which is shown on page 77, is suitable for use both as a performance aid, to help the inexperienced, and as a training aid. This, however, provides only a partial answer to our original question as to whether it is possible to write an algorithm for this subject: it is possible to provide an aid to decision making, but there is some doubt as to whether that aid is an algorithm.

One general guideline as to the use of algorithms which has come to be accepted as a principle is that algorithms should not be used to *teach* anything. Their function, it is argued, is to enable people to perform certain tasks, make certain decisions, but not to teach anything, except perhaps in that through continual use one tends to memorise the algorithm itself.

This view is, on the face of it, reasonable enough. It is possible to use an algorithm for locating faults in a car's ignition system quite successfully without understanding how a generator or a coil operates. Yet these are essential parts of the system and anyone wishing to understand the mechanics of the expected faults must understand how they work. With an algorithm this understanding is not necessary; it is all built into the structure of the algorithm itself and one has only to answer the questions to locate the faults.

This, if we accept it, implies that only those subjects are suitable for algorithmic treatment where it is possible to analyse the decision process in such a way as to isolate a complete chain of questions leading to precise outcomes.

The happy hunting ground of the algorithm writer, therefore, is seen to lie in subjects such as fault finding in machinery, human pathology, tax regulations and the law. Anything involving a weighing of various factors, a 'more or less', tends to be rejected as unsuitable for algorithmic treatment.

This approach has the virtue of being clear cut. Unfortunately it tends to be pressed too far, particularly the assumption that algorithms do not enable us to learn a subject. Certainly some algorithms can serve only to enable someone to make decisions, and tell him virtually nothing about the subject itself. Yet all algorithms must embody in some sense the logic of their subject. Why should they not display that logic too, so that if we read the algorithm through we should understand the relations between the ideas expressed?

To the purist algorithm writer the idea of reading an algorithm at all is anathema. Algorithms are meant to be *used,* not to be *read*; indeed, some, like our earlier example on looms, would make no sense at all if one attempted to read them. To him it is a matter of accident whether or not an algorithm displays the logic of its subject. What matters is simply that it should express that logic in a way that enables the user to make a quick decision. Yet why should it be a matter of indifference, whether or not an algorithm displays the logic or not?

Let us look again at the example of the Leasehold Reform Act algorithm which we worked out in Chapter 2. At one point we wish to express the provisions of the Act, which state that a freehold may be bought if (among other things) the house had a rateable value of not more than £400 if it was in the Greater London Area and not more than £200 elsewhere. Our original formulation of questions and answers for this section of the algorithm was:

Fig. 7.1

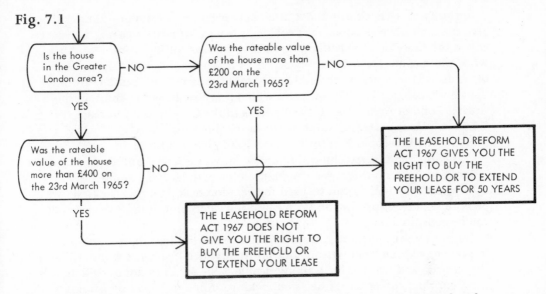

We suggested at the time that the crossover of lines was awkward and avoided it by swapping a YES and NO. This is, however, not the only way of rearranging the algorithm. Suppose we had swapped some of the questions over:

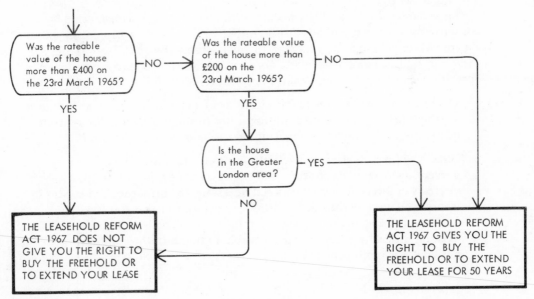

Fig. 7.2

Logically there is no difference at all between the two layouts. Each will give the same answer whatever problem it is used to solve. What is different is that the first (Fig. 7.1) *both expresses and displays* the logic of the original, while the second (Fig. 7.2) *merely expresses* it. The first sets out the rules in the order of the concepts themselves, so that we can easily translate from the algorithm back into the ideas on which it is based: 'Is the house in the Greater London Area?' (i.e. If the house is in the Greater London Area . . .) 'YES . . . Was the rateable value of the house more than £400 . . .? . . . YES . . . The Leasehold Reform Act does NOT give you the right . . .' . . . . (i.e. the limit of rateable value under the Act is £400,) and so on.

The first version, therefore, can be used for both learning and decision making, but the second can be used for decision making only. The first can be both read (for learning) and consulted (for decision making); the second can be consulted only.

For our present purpose it is sufficient to point out that there is nothing to prevent anyone from writing an algorithm which displays the logic of the subject, and using it for training people. This use does not invalidate the algorithm's other use in decision making, although it may be necessary to modify the layout in the interests of better display, perhaps by adding explanations, e.g. in our car example (page 20), explaining how the generator works. The point is not whether such adaptations are legitimate (are truly algorithms), but whether they are useful for the purpose we have in mind at the time.

This is surely the crux of the matter. What we should be considering is not what the proper use of algorithms might be, but how we can construct aids to enable people to do things in the most efficient way. In some cases these aids will be recognisable as algorithms, in others we may not feel sure ourselves what to call them. What, for example, is the following?

The problem involves an electronically controlled heat-treatment furnace in which castings are moved mechanically through all stages for heating, quenching, washing and tempering, in one continuous operation.

When the furnace alarm sounds, the correct action must be taken to prevent damage to castings or, in some cases, an explosion. In the interests of safety it was considered necessary for managers and supervisors who were not trained in furnace operations to be able to cope with such emergencies. Normal training was uneconomic and the existing manual was not comprehensible to the untrained and not designed for rapid consultation.

The following are extracts from the manual that was designed to cope with this situation. They are selected to show the sequence followed in response to the following situation:

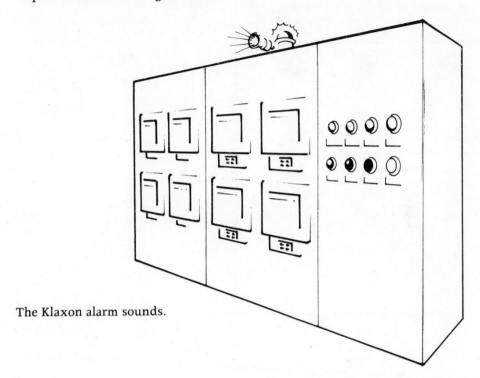

The Klaxon alarm sounds.

Page 2 of the Performance Aid Manual shows where to look for guidance.

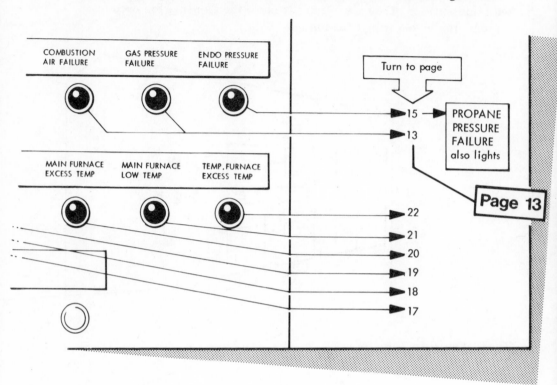

This leads the inquirer to page 13, where he is further questioned on the symptoms,

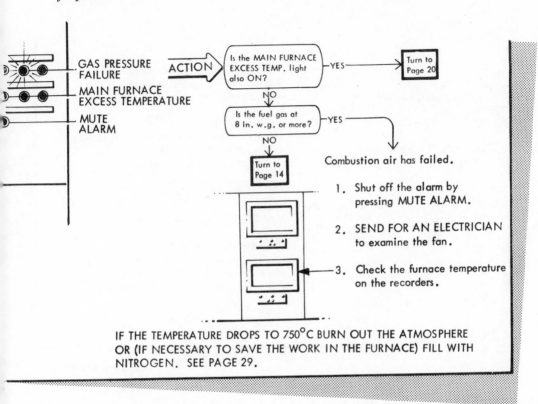

GAS PRESSURE FAILURE

MAIN FURNACE EXCESS TEMPERATURE

MUTE ALARM

ACTION

Is the MAIN FURNACE EXCESS TEMP. light also ON? —YES→ Turn to Page 20

NO

Is the fuel gas at 8 in. w.g. or more? —YES

NO

Turn to Page 14

Combustion air has failed.

1. Shut off the alarm by pressing MUTE ALARM.

2. SEND FOR AN ELECTRICIAN to examine the fan.

3. Check the furnace temperature on the recorders.

IF THE TEMPERATURE DROPS TO 750°C BURN OUT THE ATMOSPHERE OR (IF NECESSARY TO SAVE THE WORK IN THE FURNACE) FILL WITH NITROGEN. SEE PAGE 29.

and is directed to page 20, where he finds precise instructions on what to do.

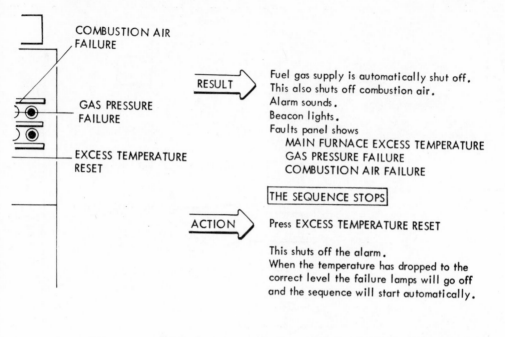

COMBUSTION AIR
FAILURE

GAS PRESSURE
FAILURE

EXCESS TEMPERATURE
RESET

RESULT → Fuel gas supply is automatically shut off.
This also shuts off combustion air.
Alarm sounds.
Beacon lights.
Faults panel shows
    MAIN FURNACE EXCESS TEMPERATURE
    GAS PRESSURE FAILURE
    COMBUSTION AIR FAILURE

THE SEQUENCE STOPS

ACTION → Press EXCESS TEMPERATURE RESET

This shuts off the alarm.
When the temperature has dropped to the
correct level the failure lamps will go off
and the sequence will start automatically.

Should we call a manual of this kind an algorithm? Its logical structure is very similar and it is certainly a 'means of reaching a decision by considering only those factors which are relevant to that particular decision'. Admittedly, the form in which it is written is far removed from anything we normally associate with algorithms, yet should we shrink from claiming it as such? Whatever we call it, here surely is the way ahead, and if in the course of adapting methods to the needs of the moment more variations arise, there is no cause for alarm, for out of these may evolve new means of communication.

8 THE USE OF ALGORITHMS – ACCEPTANCE AND FAMILIARITY

There are two main obstacles to the more extensive use of algorithms, neither of which has anything to do with the difficulty of writing them.

The first is *job mystique*. Opposition to new techniques is to be expected. What sometimes comes as a shock to the writer of algorithms is opposition by those who are at present doing the job for which he is preparing aids. This can be particularly troublesome if the same people are also his main sources of information on that job.

The reason is not far to seek. We are all accustomed to the medical practitioner's dignified silences, which we hopefully interpret as intensive consideration of our symptoms, but which, for all we know, may conceal nothing more than an unsuccessful attempt to recall what the professor had said at medical school. Whatever is occupying his thoughts, the result is always the same: he makes a decision. He decides on some treatment or none, and we cannot check his decision in any way at that moment. We have to accept the authority of the expert. One has only to imagine his reaction if we were to take an algorithm out of our pocket and start questioning his decision, to understand the whole problem of job mystique. Even if the algorithm vindicated his diagnosis, and even if it reflected his total decision-making process, we could not expect him to welcome the loss of personal prestige involved. The algorithm might represent but one limited part of his total competence, but it would still subtract from his authority by making the ability to perform that task public property.

The doctor's job involves knowledge and skill of a relatively high order, and it is probable that the job mystique that surrounds him serves largely to make his job easier by avoiding time-wasting discussion, although it no doubt has other advantages too. Unfortunately, job mystique is not by any means related to the level of knowledge and skill in the job itself; indeed, one wonders at times if any relation that exists is an inverse one. The point is that many jobs involve routine decision making based on a fixed set of logical considerations. Much legal decision making is of this kind, as is a great deal of fault diagnosis, procedural work and administration. From the point of view of efficiency there is much to be said for selecting such areas of work, turning the decision-making processes into algorithms, and employing less able people to carry out the tasks, which have now become routine, leaving the experts free to cope with the more difficult parts of the job. In certain cases this, or something like it, has proved very beneficial.

In one instance it was found that reorganisation of social welfare work had resulted in welfare officers having to cope with a much greater diversity of cases. Whereas formerly a Mental Welfare Officer, for example, had to deal only with cases within his speciality, he now had to be able to make decisions and carry out procedures in child welfare and other areas too. This involved a great deal of legislation with which he was unfamiliar, and with the increased caseload had little time to study. The solution was to design algorithms to enable the social worker to act in accordance with the legislation concerned, and thus free him for his main task of helping people.

In other cases the reaction has not been so favourable. The worker who realises with a shock that a set of algorithms is a virtual substitute for all he has learned about his job in the last twenty years is unlikely to hail the new technique as a breakthrough in communication. He is much more likely to complain that his job has been 'deskilled', and one can feel a good deal of sympathy for his attitude, if not for his conclusion.

Sympathy apart, what can we do about the problem? Once the typewriter had been invented no one used quill pens any more in the interests of efficiency; and once the algorithm has been invented it is irrational not to use it wherever it is appropriate. Of course, the quill pen is more artistic than the typewriter; it requires greater skill, specialised knowledge and a whole set of equipment and techniques. Did the typewriter 'deskill' the specialist in copperplate handwriting? – or did it free him for tasks more suited to his ability?

One final warning, however: it is important for us as writers of algorithms to distinguish between the resistance that springs from a threat to mystique, and a genuine objection based on recognition of greater complexity than we have allowed for in our logical mapping. We disregard this at our peril, for we are always at the mercy of hard fact. Either our algorithms get results or they do not, and if they do not, no mystique of ours will save us.

On the subject of job mystique among writers of algorithms, perhaps we should add at this point that in our view it can easily be overdone. It is also unnecessary. The subject has developed considerably in recent years, both in theory and practice. This book, for example, which has a heavy practical bias, contains more material than a casual enquirer would wish to master in detail. That material is of value, however, only in so far as the writer of algorithms is able to make use of it.

The only reputation we, as writers of algorithms, can reasonably covet is for ingenuity in devising solutions to communications problems and for utter reliability in putting them into practice. This demands a good deal of hard work, but without it all our inspiration and ingenuity count for nothing. The test is what happens when our first reader uses one of our materials to solve his problem, and that is the only test that matters.

It is not only those who feel that their job mystique is threatened who object to using algorithms to solve problems. Even those who have no vested interest and who admit that an algorithm has enabled them to solve a seemingly insoluble problem may nevertheless treat algorithms with a good deal of reserve. What are the causes of this? How can they be overcome? Just what are the limits of algorithms in actual use?

The following observations are based partly on a small study of the use of algorithms as means of enabling people to interpret a piece of legislation, and partly on extensive experience of their use in many fields.

First, are there people who cannot use algorithms at all? The main limitation is clearly the ability to read the questions, including the ability to relate whatever technical concepts these contain to the job in hand. If one cannot do this, then no written materials of any kind are of any use. The question is whether a level of intelligence, or a spectrum of abilities, exists which allows some people to use conventional prose descriptions more effectively than algorithms for problem solving. We have found no-one in such a category, although we have found a number of peripheral problems.

Some people, for example, have a tendency to read across the top of the algorithm as if it were a line of type, paying no attention to the YESs and NOs. Others seem to hate diagrams. For such people other presentations, such as the list structure, are suitable. In practice, however, it appears that the main objections to algorithms come from people who are able to use them to solve problems but have certain misgivings about them. What then are these misgivings?

Algorithms do not help you to *understand* the problem,

We have considered this question from the point of view of the writer of algorithms in the previous chapter: algorithms may or may not be designed to display the logic of a subject as well as to enable the user to reach decisions on it. Even if they are not, however, is the objection justified?

F

After all, there is nothing to prevent the user from taking steps to understand the subject in depth, while using the algorithm to prevent himself from making mistakes in the meantime. Besides, many people do not have the ability and many more do not have the time or opportunity for protracted study; an algorithm can be of great value to such people in allowing them to do the job, whether they understand the reasons for what they are doing or not.

'How do we know that the designer of the algorithm has got it right?'

How do we know that *anyone* who purveys information in any form has got it right? After all, there have been mistakes even in Acts of Parliament. Behind the question, however, lies a valid point, if not an objection. Algorithms must be used with discretion. We cannot reasonably expect more from them than they claim to do. We may, for example, use an algorithm to find out whether or not we are liable for capital gains tax, but we cannot expect to find out from any algorithm what our best defence would be against a particular charge of defrauding the Inland Revenue. To formulate such a defence would involve assessing not only the law but the circumstances of the case, the results of similar cases in the past, the personality of the judge, and a good deal more besides. There is no reason why algorithms should not include judgements or estimates, but a decision based almost entirely on estimates in a fluid situation is not suitable for algorithmic treatment.

'An algorithm gives you no opportunity to check whether it is right or not, whereas a conventional text does.'

This depends on the conventional text, in particular on how complete it is. If the text is an Act of Parliament, for example, then the objection is valid, theoretically at least. All the information is there, so we have only to read, compare and draw our own conclusions. In practice this can be a very tedious process, involving references to other legislation, comparing sections and subsections of the same Act together, and deciding what the significance of any variations is. Explanations such as the following must be coped with as a matter of routine:

1. In the case of a rebate application by such a person in respect of such a hereditament as is mentioned in paragraph (a) of section 5(3) of this Act, the applicant's reckonable rates shall, subject to the provisions of this section, be:
(a) the amount of the rates chargeable on that person in respect of that
     hereditament for the rebate period to which the application relates, less

(b) the proportion of that amount which, by virtue of subsection (3) and apart from subsection (4) of this section, is or would be the reckonable rates in relation to that rebate period of any person or persons entitled to apply for a rebate in respect of any part of that hereditament by virtue of paragraph (c) of the said section 5(3).

Add to this the virtual obligation to read every word of the Act to make sure whether or not it applies to the case in hand, and for most of us there would seem to be little point in making the attempt at all.

Transforming an Act into flow charts and algorithms brings all the material together and specifies the decisions to be made. For the expert, of course, this process has no advantages; he has already sorted the Act into the same categories in his own mind. For the non-expert, there are three possibilities: he can grapple with the Act, if he has the time and the mental equipment; he can pay an expert to explain it to him; or he can take what steps are available to him to ensure that the algorithms are accurate and then use them. This assurance should, of course, be provided by the writer of the algorithms.

On this last point we would emphasise that an algorithm which has not been checked by an expert in the subject and validated with a sample of potential users is a snare and a delusion. Unfortunately, intensive study, such as is necessary to produce algorithms, leading to an ingenious solution to a problem in communication, is only too likely to generate a feeling of infallibility. We have, we feel, thoroughly sorted this subject out; it is crystal clear in our minds, so surely it must be right. With due respect to Descartes, however, what we take to be the clarity and distinctness of our ideas is often bought at the expense of serious omissions from the data on which they are based. The remedy is obvious: we must get our work checked and validated, and having done so we owe it to the reader to tell him that we have taken these steps to guarantee the information we are giving him.

'People are conservative'

Finally, one obstacle to the use of algorithms is often met with but is rarely given any overt expression. It is revealed only as an attitude; many people react adversely to this new and strange method of communicating with them. It is tempting to ascribe this simply to the natural conservatism of the human species. To do so may give us a warm glow as we find ourselves aligned with progress, but the temptation should be resisted, for two reasons.

The first is that the attitude has more substance than this. When we meet a new method of communication we immediately try to relate it to other methods of communication we already know. Conventional prose gives many clues as to the attitude and intention of the writer. Through experience of it we have acquired the ability to detect a note of resentment or patronage in the writing; to distinguish enthusiasm from mere reporting; we can, we say, 'read between the lines'. When we are confronted with an algorithm for the first time, we realise that we cannot interpret it in the same way, but we have not yet acquired the ability to evaluate it in its own terms.

The second reason why we should not dismiss negative attitudes is that they are part of our problem in communication. However skilfully we have designed our algorithm, it is futile if no one uses it. How can we get people who react adversely to it, to use the aids we are preparing for them and which it is manifestly in their interests to use?

A certain amount of light was thrown on this problem by a recent pilot study on the value of algorithms as a means of explaining a piece of legislation. The purpose of the study was to compare the effectiveness of a conventional question-and-answer leaflet, as issued by the Ministry concerned, with a flow-chart algorithm and a list-structure algorithm, both presenting exactly the same information. The sample was small, but the indications were that the flow-chart algorithm won hands down on effectiveness, provided that the use of the algorithm was preceded by a brief familiarisation. Once people had answered one question and followed the correct line to the next box they could use the algorithm.

The list structure was also effective and was perhaps more popular than the flow-chart algorithm, presumably because it did not look so unfamiliar. The leaflet was almost a non-starter.

The fact that so little explanation of the algorithm was given, but that that little was necessary in many cases, may point the way ahead. Nothing is so encouraging as success, and the solution of a problem which faces us by means of an algorithm is an impressive experience. What is particularly impressive about it is the ease and speed with which we can arrive at a solution. People who have once experienced this are likely to be willing to use algorithms to get the answer in future, if only they have confidence that the answer the algorithm gives them is the right one.

What we are saying, therefore, in effect is that algorithms need to become respectable. The user should be able to feel assured that the simplicity and speed have not been bought at the expense of accuracy. The remedy is clear: it should already be a matter of routine to have one's algorithm verified by an expert. That expert, if he is worth his salt, should be prepared to stand by his decision publicly. We would then be in a position to give the user of our algorithms the assurance he needs. Perhaps it should become fashionable to include a kind of 'hallmark' with algorithms—a small panel stating that the algorithm has been checked and approved by a named expert.

Logically, there is everything to be said in favour of this approach; practically, you have first to catch your expert.

9 LAYOUT AND VISUAL DESIGN

Perhaps the most obvious general point to make about the layout of algorithms is that they do not necessarily fit neatly into the format of normal pages. In conventional book design, words, sentences and paragraphs, with an occasional diagram, are made up into blocks of type in such a way as to promote the legibility of the text and to form a pleasing pattern. Algorithms – and it is particularly the flow chart type we are considering here, since these cause most of the problems – are difficult to fit into a page format in this way. Any attempt to fit them in, either in the interests of saving space or in the interests of visual appeal alone, is likely to make them needlessly difficult to use.

In the following, for example, the questions and outcome boxes appear to have been written into the space available and joined together with lines afterwards. The one virtue of the layout is that it fits the page format.

TO ELIMINATE HUM

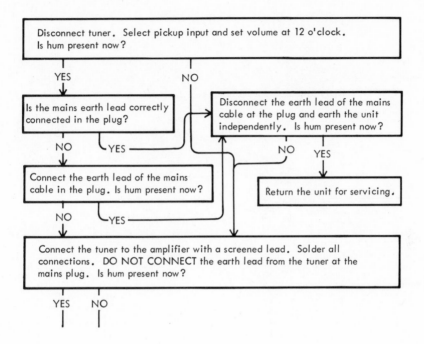

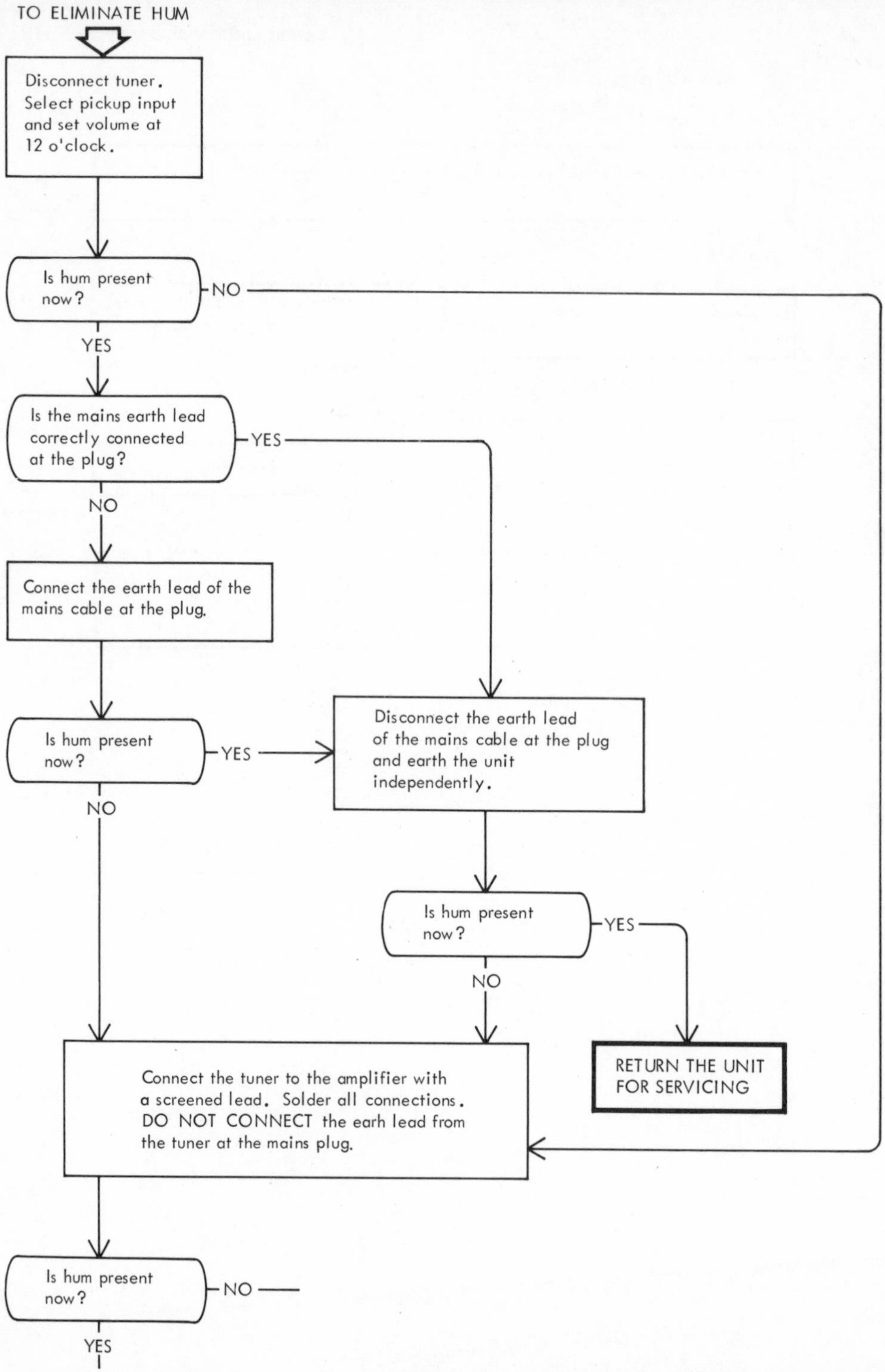

TO ELIMINATE HUM

Disconnect tuner.
Select pickup input
and set volume at
12 o'clock.

Is hum present
now? — NO

YES

Is the mains earth lead
correctly connected
at the plug? — YES

NO

Connect the earth lead of the
mains cable at the plug.

Is hum present
now? — YES → Disconnect the earth lead
of the mains cable at the plug
and earth the unit
independently.

NO

Is hum present
now? — YES

NO

RETURN THE UNIT
FOR SERVICING

Connect the tuner to the amplifier with
a screened lead. Solder all connections.
DO NOT CONNECT the earh lead from
the tuner at the mains plug.

Is hum present
now? — NO

YES

In the revised version of the same algorithm, (Figure 9.2), two things have been done. Firstly, the boxes have been so positioned as to avoid the crossover of lines and to give a general movement of the algorithm from top to bottom and from left to right. Secondly, the questions ('Is hum present now?') have been taken out of the boxes giving instructions and placed in boxes of their own. It can hardly be denied that the second version is a great deal clearer. It also takes up more space, but then which costs more – paper or problem solving?

Algorithms may also be ill-designed visually for precisely the opposite reasons. A good deal of attention may be given to the finished appearance and there may be no penny-pinching on space, but the result can still be difficult to use. The placing of the YESs and NO*s* in Figure 9.3 can surely not have been decided on for ease of comprehension. Presumably they have been placed so far away from the questions which they answer as part of an overall visual design; unfortunately the visual designer has not appreciated their purpose, and his layout is merely decorative rather than functional.

Sometimes an attempt is made to do away with arrows altogether. After all, if the line we are to follow from a given box is indicated by the YES or NO we have only to follow that line until it reaches another box. What is the point of arrows at all? The point is surely that the arrows are confirmatory signs. They give the user assurance and so enable him to use the algorithm more quickly. In Figure 9.4 the presence of arrows, or better placing of the lines, would have avoided the risk of confusion at the area indicated.

There are many possible faults in the layout of algorithms and we cannot hope to describe them all. What we wish to stress here is that layouts must be practical. If they can also be elegant, so much the better, but the elegance must never override the practicality. Figure 9.5 shows a layout in which a neat appearance has been allowed to do this.

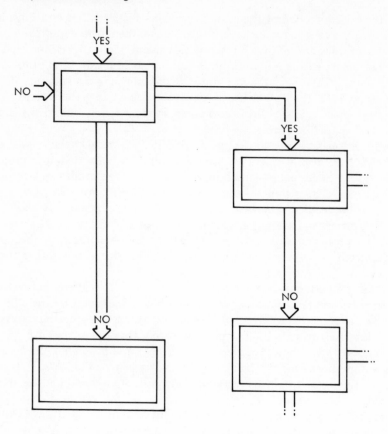

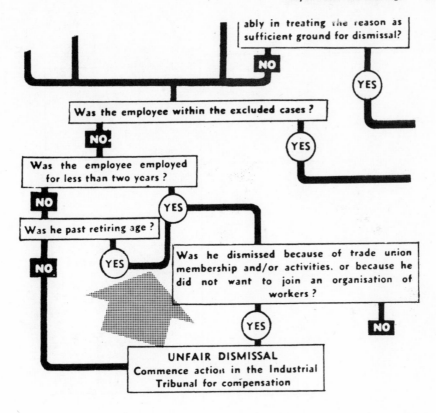

ably in treating the reason as sufficient ground for dismissal?

NO

YES

Was the employee within the excluded cases ?

NO   **YES**

Was the employee employed for less than two years ?

NO   **YES**

Was he past retiring age ?

NO   **YES**

Was he dismissed because of trade union membership and/or activities. or because he did not want to join an organisation of workers ?

YES   **NO**

UNFAIR DISMISSAL
Commence action in the Industrial Tribunal for compensation

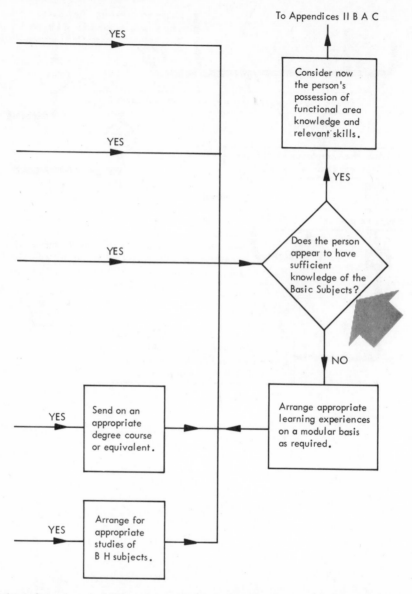

To Appendices II B A C

Consider now the person's possession of functional area knowledge and relevant skills.

YES

Does the person appear to have sufficient knowledge of the Basic Subjects?

NO

YES

YES

YES

Send on an appropriate degree course or equivalent.

YES

Arrange appropriate learning experiences on a modular basis as required.

YES

Arrange for appropriate studies of B H subjects.

Which line do we follow if we answer NO to the question marked? It is by no means obvious at first glance that we are led back through the instruction box to the same question box again, and it is the function of a good layout to make such things obvious, or people will be distracted from the task of using the algorithm to puzzle over the mechanics of the presentation. Whether it is helpful to build a sequence like this into an algorithm at all is another matter, but if we do we should make it clear how the reader is to use it.

So much for what should not be done in designing algorithms. What makes for good visual design?

The essentials are quite simple. Firstly we have to make it as easy as possible for the user to choose the right line according to what his answer is at each box, and secondly we have to avoid throwing him off the track by unnecessarily tangling the lines.

The main principle concerning the box itself is, as we suggested earlier, to bring the YES and NO as close as possible to the question. In earlier chapters we have consistently used the layout:

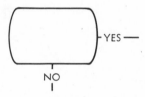

Many variations are possible:

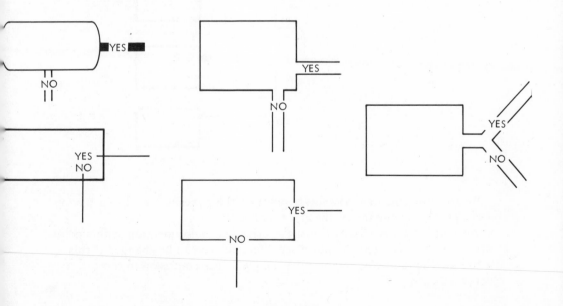

Unfortunately, some of these take more time to draw and are therefore more expensive to produce, although they may be justified in special circumstances. For example, the problem we mentioned in the last chapter, where some people read across the top of the algorithm paying no attention to the YESs and NOs, can be tackled in terms of layout.

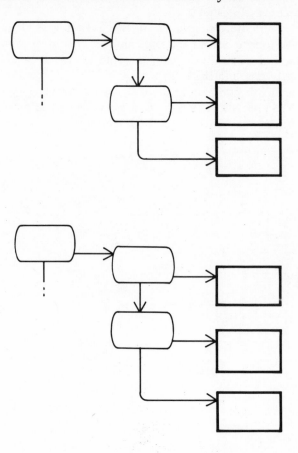

By attaching the lines at a lower point on the boxes we can avoid having a straight line at the top of the algorithm at all.

Although it is not strictly a visual matter, we might mention at this point that even the seemingly standardised YES and NO can be changed if this is useful. The following might, for example, cause confusion to some people:

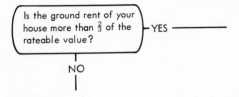

Part of the difficulty can be removed by confirming the answers, thus:

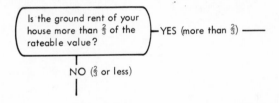

There is not much to be said about the arrows in an algorithm, except that they should not be too prominent. We have seen examples in which the arrows were so over-emphasised as to appear more prominent than the questions themselves. As far as lines are concerned, a distinction is sometimes made between those which denote replies to questions and those which run from an instruction to a question, thus:

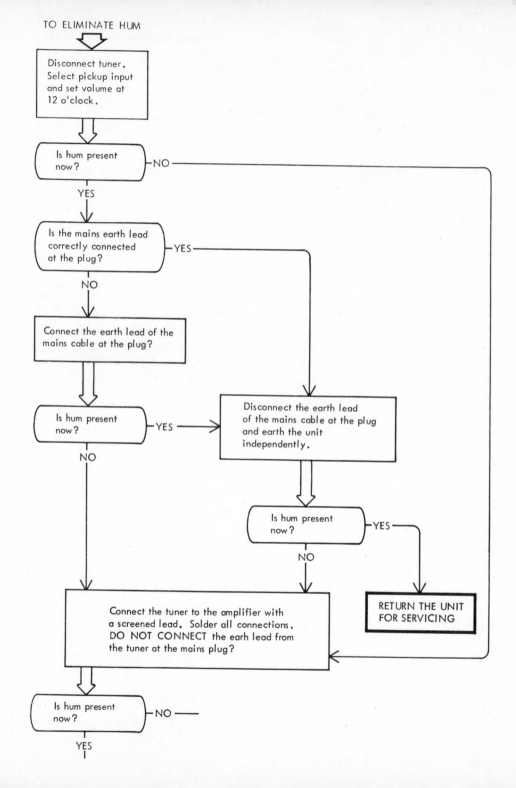

We have used these 'procedural' lines ourselves, particularly in algorithms on legislation; they seem to make for greater clarity, although we cannot claim to have any objective proof of this as yet.

Our own general style of layout, which we have used throughout this book, is based on apparent ease of use and on speed of drawing. If algorithms are to be published, the latter should not be neglected; if it is, costs can mount rapidly, particularly where alterations to artwork are involved. Draughtsmen and printers are not yet accustomed to algorithms and a few simple standards can be a great help.

The general style we have used includes:

    large arrow for 'start'
    round-ended boxes for questions
    heavy square-ended boxes for outcomes
    light square-ended boxes for instructions
    YES and NO close to the box they answer
    general movement of the algorithm from left to right and from top
        to bottom
    avoidance of crossovers where possible.

Other refinements may be introduced, such as the 'procedural' double arrows we mentioned earlier. In certain cases, too, it may be possible, without distorting the algorithm, to run all the YES lines in one direction and all the NO lines in another.

We would stress that every algorithm is an individual case. As with any piece of writing, there are certain rules which make for ease of understanding; these are normally applied by the writer. There are also certain other rules which make for legibility, and these in conventional writing are applied by the designer and printer. In the case of algorithms, the writer must concern himself with both, since the spatial design expresses an important part of the meaning. It is a commonplace of writing that the good writer knows how to bend the rules to his advantage. One of the aims of this book is to help the designer of algorithms to acquire this ability.

INDEX